300 OF THE
MOST ASKED QUESTIONS
ABOUT ORGANIC GARDENING

Answered by
The Editors of Organic Gardening
and Farming® Magazine

Compiled and Edited by
Charles Gerras

With
Joan Bingham, Joanne Moyer and Irene Somishka

Special Editorial Assistance by
Betty Frederick

T.M. REG

Book design by Karl Manahan
Jacket design by David Traub

RODALE PRESS Book Division, Emmaus, Pa. 18049

Standard Book Number 0-87857-045-4

Library of Congress Catalogue Card Number 72-84796

PRINTED IN U. S. A.
OB — 358

Third Printing — July 1974

Printed on Recycled Paper

TABLE OF CONTENTS

PAGE

CHAPTER 1. INTRODUCTION.................... 1

CHAPTER 2. SOIL............................. 6

How to Have Fertile Soil...................... 6
Crop Rotation............................... 15
Aeration.................................... 16
pH (Acid or Alkaline?)...................... 17
Microorganisms............................. 20
Humus..................................... 23
Erosion.................................... 24

CHAPTER 3. FERTILIZERS..................... 27

Chemical vs. Organic........................ 27
Nitrogen, Phosphorus, Potassium............ 32
Manure..................................... 36
Rock Fertilizers............................. 39
Trace Elements............................. 40
Taboos..................................... 41
Miscellaneous.............................. 43

CHAPTER 4. COMPOST........................ 45

Importance of Composting................... 45
Material.................................... 52
Compost Heaps and Containers.............. 59
Application................................. 64
Do's and Don'ts............................ 68
Miscellaneous.............................. 70

CHAPTER 5. EARTHWORMS.................... 72

CHAPTER 6. WEEDS, SEEDS AND SEEDLINGS.... 77

Eliminating Weeds.......................... 78
Seeds and Seedlings........................ 81

CHAPTER 7. TREES AND SHRUBS.............. 87

CHAPTER 8. MULCHING....................... 100

Pros and Cons.............................. 100

TABLE OF CONTENTS

PAGE

Materials. 103
Miscellaneous. 107

CHAPTER 9. VEGETABLES. 109

Planting. 109
Vegetables—Problems. 114
Fertilizing Vegetables. 118

CHAPTER 10. LAWNS. 121

CHAPTER 11. INSECTS. 142

Good and Bad Bugs. 142
Controlling Pests with Plants. 156
Biological Control. 158
Man and Beast. 159
Protecting Trees. 161
Pesticides. 164
Miscellaneous. 166

CHAPTER 12. ANIMAL PESTS AND
　　　　　　　THEIR CONTROL. 168

CHAPTER 13. THE BIRDS & THE BEES. 173

Birds. 173
Bees. 176

CHAPTER 14. PLANTS. 179

House Plants. 179
Roses. 186
Annual, Biennial, Perennial. 190
Miscellaneous. 194

CHAPTER 1

INTRODUCTION

Gardening or farming organically means maintaining a fertile soil by applying nature's own law for replenishing it—that is, the addition and preservation of humus. We use organic matter instead of chemical fertilizers and, of course, we make compost and we mulch. With time and the uncompromising practice of the organic methods, increased yields will come.

Organic gardeners believe that the balance of nature must be respected. Each part has its own sphere of activity and each of these fuses with and complements other related parts. So, a foreign substance that alters one part may affect half a dozen others—most often to our disadvantage. Spraying that kills pests also kills beneficial insects and birds, contaminates the soil, leaves poisonous residues on foods grown for man and animals and is a financial burden to the farmer or gardener.

The soil is a storehouse of living organisms which must be fed and cared for as any others. Bacteria, fungi, insects and earthworms inhabit it by the millions, using organic matter for food and in turn preparing it for living plants. Concentrated chemicals, on the other hand, cannot be continually added to the soil to destroy harmful insects and disease organisms without harming the needed beneficial microorganisms.

The organic gardener believes that all suitable organic matter should go back to the land, either by mulching, spreading it over the ground to protect growing things as it decays itself, or composting (mixing organic substances and encouraging the mixture to decay for eventual return to the soil as a fertilizing aid.) Only those materials which ordinarily are not applied to the soil (garbage, for example) need be composted.

Besides compost made of plant matter, the organiculturist usually employs as fertilizers such substances as raw phosphate rock, dolomite, ground oyster shells and miscellaneous ground rocks, such as granite dusts and pulverized limestone.

The use of poison sprays and dusts in orchards and on crops is taboo, for there is definite evidence that the strengthening of a plant or tree by the use of compost makes the plant or tree much less susceptible to infestation by insects or disease than does recourse to sprays. The organiculturist believes that, even on a commercial scale, composting can eventually eliminate dependence upon poison sprays.

It is true that some natural insect controls may be necessary until, and possibly after, the gardener has been practicing the organic method for several years. The soil must become fertile and insect parasites and predators must be encouraged. Exterminating all harmful insects is not the goal of the organic method. Good yields, truly safe food and sensible insect controls are the answer.

The first pioneer of organic methods was Sir Albert Howard. Although he was not the earliest critic of the ills of modern agriculture, he was the first to offer a remedy. He saw the solution to the problem in the thorough study and application of methods evolved by nature. Sir Albert deplored the unnatural separation between the soil, crops, livestock and human beings. He believed these to be all part of a natural complex, and thought research on each one without reference to the others was dangerous. During the early 1900's many acute observations, backed up by thousands of field trials, went into this work and Sir Albert succeeded in proving that a soil well supplied with organic matter grows bigger yields of healthier crops.

In 1940, J. I. Rodale, father of organic gardening in America, read of the work of Sir Albert Howard. His teachings had such a profound effect on Mr. Rodale that he immediately purchased a farm in order to grow food organically for himself and his family. When he saw the improvement in the health of all the Rodales he became anxious to share his success with others. To accomplish this he founded Organic Gardening and Farming Magazine, and later, the Soil and Health Foundation, a nonprofit organization which supports vital agricultural research.

Another prestigious leader in the organic movement, Britian's Lady Eve Balfour, authored the valuable book, "The Living Soil." In 1946, she became the first president of England's Soil Association—a group that has done much important research into the organic method.

A pioneer of municipal composting in the United States, Dr. G. H. Earp-Thomas, developed his principle of the "continuous-flow digester," an invaluable composting device.

To give credit to all who have devoted their lives to the organic movement would be impossible. Suffice to say that their contributions have been great.

The term "organic" can be misleading. For instance, "organic phosphate" is a chemical, as are many things labeled "organic." Chemically "organic" describes any compound which contains carbon as an essential ingredient. While biologically organic, (which is what we mean by organic) pertains to any substance which is alive or that was once alive. The only known substances that are alive are in plants, animals and man.

The bio-dynamic method of farming is a highly-refined technique. While most organic gardeners don't differentiate between compost of varying materials, the bio-dynamic gardener mixes carefully portioned amounts of certain raw materials to form compost according to specified formulas. The bio-dynamicist is intent on producing compost in such a way as to keep as many of the original nutritional elements as possible.

Switching from a chemical garden to an organic one can be done in as little time as it takes to collect organic materials and apply them to the soil. The main job at the start is to round up plenty of organic fertilizer material. Some of these materials, such as commercial compost mixtures, dried manures, etc., can be worked directly into your soil to increase its humus content. Another portion of these materials, such as hay, sawdust, cocoa hulls, etc., should be set aside for later use as mulches.

Plan on setting up a small composting area immediately, using the remainder of the materials you've collected. In less than a month's time, you can have a high-grade organic compost that will be the real start of your organic garden.

Commercial organic fertilizers are also available at just about any garden supply store. So there's no reason to postpone converting to the organic method until the gardener gathers these materials on his own. Any gardener will be pleased with the noticeable improvement in his soil brought about by organic gardening methods.

The organic method is just as practical on a large farm as it is in a small garden. (One need only mention Friend Sykes, whose 750-acre farm, Chantry, is run entirely on the organic method.) Briefly, the additions of farm manures, rock powders, green manures, turned under crop residues, a rotation plan and careful soil management are the keys to organic farming. In most farm crops, only a small portion of the total plant is used for food. If the remainder of the crop is returned to the soil, along with rock powders and other easily available organic materials, crops will not suffer from nutrient deficiencies.

Proper nutrition is an integral part of the organic

method of gardening and farming. Aside from the many other advantages of the organic method, the sound nutrition that is achieved from eating naturally grown foods is reason enough to follow the organic method. The relationship between soil and health is as basic as the very processes of nature, the axiom "you are what you eat" assumes greater and greater impact as experiments show the wonderful results of eating carefully selected, organically grown foods.

The following questions were posed by organic farmers, organic gardeners and would-be organic gardeners who have come to look to Rodale Press as new mothers still look to Dr. Spock, for reassurance, information and advice. The problems range from the very basic to the highly specialized, from how to get rid of the wrong bugs to how to encourage the right ones, from how to raise trees to how to kill weeds.

The seasoned organic gardeners are sure to come across some worthwhile nuggets of new information in the pages that follow; for those new to organic gardening, this book may seem like the answer to a prayer.

CHAPTER 2

SOIL

Here's where it all begins. Soil contains the raw material that yields a beautiful lawn, a graceful tree, a juicy tomato, or a velvety rose. Fledgling gardeners soon learn that fertile soil is no accident; it must be encouraged. A gardener must know the strengths and weaknesses of the land; then he must learn how to take advantage of the former and compensate for the latter. Tired-out land can be revitalized; cement-like layers known as "hardpan" can be treated. There are natural answers to most of the soil problems that seem unanswerable.

Know the capacities of your soil thoroughly so you can avoid frustrations and enjoy the satisfaction that comes with growing bumper crops on land that is particularly suited to them.

How to have Fertile Soil

I

Q. What conditions tend to make the soil unproductive?

A. The classic undesirable growing conditions of a soil are: the presence of a hardpan; an extremely sandy, or extremely clay-like state; and an excessively alkaline or acid status.

Hardpans are impervious horizontal layers in the soil that may exist anywhere from six inches to about two feet below the surface. A true hardpan is formed by the cementing together of the soil grains into a hardstone-like mass which is impervious to water.

A more common condition is an impervious layer in the subsoil caused by the pore spaces becoming filled with fine clay particles. Such "tight clay" subsoils, called claypans, are generally associated with an extremely acid condition so that they are objectionable, both physically and chemically.

Q. How does hardpan hurt the soil?

A. When hardpan exists, the surface soil is cut off from the subsoil; no new minerals are added to the lower part of the soil; plant roots are often unable to penetrate these layers. Plant roots usually grow down to this hard layer and then extend horizontally along the top of it. This results in shallow-rooted plants which may suffer from lack of nutrients otherwise available in the subsoil and from lack of water during the dry summer months. Often such shallow-rooted plants die out completely from lack of water during dry periods while plants nearby, where there is no hardpan, flourish and grow vigorously.

Q. How can hardpan be remedied?

A. The best and most universally used method of breaking up hardpans is by subsoiling, a kind of plowing that involves cutting into the soil with a subsoil chisel or "killifer" to a depth of about 16 to 30 inches. In exceptional cases the chisel may go to a depth of over five feet. The power requirements for these heavy, deep-working tools are obviously high. Sometimes the combined power of three of the biggest track-layer tractors is needed. For the lighter tools with a single chisel, penetrating up to 30 inches deep, the power of the ordinary wheeled two-to-three plow farm tractor is usually sufficient.

Other common methods for overcoming a hardpan are incorporating large amounts of organic matter into the soil to make it more loose and friable, planting deep-rooted cover crops, improving soil aeration, and encouraging earthworms. With these methods, however, it may take several years to do the job, which is why subsoiling is the most frequently recommended method.

Q. How can a home gardener fight hardpan on a small scale and without investing in elaborate equipment?

A. Here is how one gardener, Fred L. Christen, describes his attack on the problem of hardpan in a small (6 by 20 feet) plot in the corner of his backyard: "For $5.95 I bought a 6-inch posthole auger. I planted five pepper plants and five tomato plants. With the auger I dug a hole about two feet deep for each plant. This was well into, if not through, the hardest packed soil. Each hole was filled with a mixture of compost and uncomposted material. It was watered down to drive out the air and then a plant was set right on top of that 'organic core.' Only organic material was used to fill the hole. All packed soil removed with the auger was thrown on the compost pile for the worms to aerate.

"A two-foot hole can be dug in about three minutes and it will hold about five gallons of organic material, putting it where it will do the most good. With alternate placement of holes over a period of years, an entire garden could be enriched to great depths. This will combat the hardpan problem and give plants the deep root system necessary to withstand heat, wind and insects."

Q. What is a good cover crop to build soil in the home garden?

A. One suggested winter cover is a seed mixture of one pound of rye grass and two pounds of crimson clover for each 1,000 square feet of garden space. Sow from mid-August to mid-September. Winter oats, barley or rye, seeded at a rate of 4 to 6 pounds to 1,000 square feet, may make better growth and cover when seeded late. If fall and winter vegetables remain in the garden, seed cover crop between the

8

Make auger hole.
Dig down 24 inches.
Check depth with yardstick.
Fill hole with organic compost.
Mulch around finished planting.

rows. Plow under winter cover crops in the spring, usually by May 1, before they make much growth.

Q. What adverse effects do clay and sandy soils have?

A. The particles in an extremely sandy soil are comparatively large, permitting water to enter the soil and pass through it so quickly that it dries out very rapidly. This rapidly moving water also carries valuable nutrients with it. At the opposite extreme, a very clayey soil has particles so fine that it tends to compact, which makes cultivation difficult and interferes with the oxygen supply for the plant roots. Water can do little to enter the impervious clay soil and runoff is very common during rainfalls.

Q. What can be done to eradicate both sandy and clay soils?

A. The solution for both of these problem soils is the same—the addition of vast amounts of organic matter. Spade-in compost, use a mulch and spread raw organic matter on the soil and spade it in. Cover crops and green manures which are later turned under are also beneficial.

Q. What causes soil to become alkaline?

A. An alkaline soil results from one of two conditions: (1) an accumulation of soluble salts, usually the chlorides and sulfates of sodium, calcium, magnesium and sometimes potassium; (2) large amounts of absorbed sodium that is either toxic to plants directly, or harmful to them by making the soil impermeable by water.

Q. What happens to plants in an alkaline environment?

A. The strong soil solution has a corrosive effect on the roots and stems causing an actual shrinking of the tissues, making them less able to take up water. The large amounts of salts in the soil displace needed nutrients, causing nutritional deficiencies, especially of phosphorus, iron and manganese.

Sodium causes a breakdown of the soil humus and clay. This results in the loss of good soil struc-

10

ture and the beginning of puddling and erosion. Water infiltration and root penetration are hindered and aeration is reduced. The anaerobic conditions thus produced make toxic compounds of many of the elements present. Microbial life ceases and the soil is dead.

Q. How can alkaline soils be reclaimed?

A. Drainage is very important for reclaiming alkaline soils, especially where the water table is high. The soil should be permeable to a depth well below the root zone. The ground water level should never come nearer than ten feet below the surface.

Once adequate drainage is established, leaching will help remove the salts. Here is an important point: whenever possible, irrigation water should be applied well in excess of the amount needed by the crop. And if the irrigation water itself contains harmful salts to any degree, apply extra-large quantities to make sure these salts, too, wash down. Each field or garden area should be carefully leveled so the water will enter the soil uniformly.

There are a number of other ways to overcome an alkaline condition. Acid peats can be purchased in carload lots. Or you can collect waste organic materials and apply them as a mulch.

Manure or sludge (20 tons or more to the acre) is also excellent. Growing green manure crops such as sweet clover, and plowing them in increases the carbon dioxide content of the soil air, causing displacement of the harmful sodium by calcium.

Other good alkali-resistant crops are sugar beets, cotton, rye, barley and sorghum. Alfalfa is very good because it requires a flooding-type of irrigation that aids leaching, and its tough roots help break up the packed soil. Always prepare a good seedbed, have the soil moist to insure germination and use more seed than usual.

Clovers do well in the Gulf Coast states.

Austrian winter peas add nitrogen to the soil. CREDIT U.S.D.A.

Bur clover nitrogen nodules grown in Pinto clay loam soil.

Q. What happens when soil is acid?

A. Extremely acid soil means that the bacteria which decompose organic matter cannot live. Manganese and aluminum are so soluble in very acid soil that they may be present in amounts toxic to plants. Yet strong acidity decreases total nutrient availability and plants may literally starve to death for one essential mineral while they have so much of another that it poisons them.

Q. How can acid soil be neutralized?

A. If a test indicates your soil is too acid, use natural ground limestone or materials like wood ashes, bone meal, dolomite, crushed marble and oyster shells, all of which contain liberal amounts of lime.

Q. How do legumes help build soil fertility?

A. Leguminous crops are active nitrogen gatherers. They furnish more nitrogen to our crops than farm manures and fertilizers combined. This comes primarily from the air through a process called nitrogen fixation. Certain bacteria attach themselves to the legumes' roots and form nodules which are rich in nitrogen. Leguminous crops are therefore adding extra fertility to the soil because they encourage bacteria which make atmosphere nitrogen available for plants in the soil. As they sink their roots down, legumes not only aerate the soil, but add valuable organic matter to it in addition to bringing minerals to the surface.

Q. What are some of the popular legumes?

A. Alfalfa is one of the most practical and widely grown legumes. Others include alsike and crimson clover, hairy vetch, lespedeza, and burr clover. Such plants as soybeans, cowpeas, velvet and field peas, snap beans, and peanuts—although legumes—are not usually grown for green manuring purposes because they do not develop deep root systems as do the others. Like all legumes, however, they gather nitrogen and their yields and quality are increased

13

by inoculation. It is always best to consult your local county agent to learn which variety is best suited to your area.

Q. How can the organic method help improve the structure of the soil?

A. The organic method improves soil structure by increasing its organic matter content. There are four general ways to do this:

   a) mulch regularly with organic materials, working them into the soil after each crop;

   b) turn under cover crops;

   c) work surface plant residues, additional materials with fertilizer and manures into the soil where they will decompose—this is known as sheet composting;

   d) make a compost heap and work in finished compost.

Q. What effect does organic matter have on soil that makes it so valuable?

A. Organic matter promotes a granular structure which permits soil to hold more of both water and air. This change brought about by added organic matter may mean:

   a) A more extensive plant root system.

   b) More water entering the soil faster.

   c) Less water flowing from the land and thus less erosion.

   d) Greater aeration.

   e) Less blowing of the soil due to a more moist surface.

   f) A greater amount of water stored in the soil for use by plants.

   g) Less soil baking and less crust formation.

   An increase in soil organic matter keeps the soil at a more uniform temperature.

   Soils high in organic matter, especially when it is used as a surface mulch, lose less water by evaporation.

Soluble plant nutrients which may otherwise leach out of reach of plant roots are held in place by partly decomposed organic matter (humus).

Decomposing organic matter is to growing plants what self feeders are to livestock. The matter aids directly in promoting the proper amounts of all the factors of plant growth except sunshine.

Another function of organic matter in the soil consists of furnishing to growing plants certain growth-promoting substances. Soils high in organic matter are healthier for plants and animals, as well as for man.

Crop Rotation

II

Q. What is meant by crop rotation?

A. Crop rotation is a regular scheme of planting whereby different demands are made on the soil each year. In contrast to the continued growing of one or more soil depleting crops on the same land, or to irregular cropping of the land without a definite plan, systematic rotation has many advantages.

Q. How can crop rotation be used in farming?

A. A four-year rotation might be in corn, oats, wheat and clover in four fields of similar size. Each year the farmer has all four crops, but in different places, rotating them clockwise. The fifth year corn goes in where it was at first and the repetition starts. Depending on his needs, he may have a three- or five-year rotation, or even a larger one, sometimes leaving a perennial like alfalfa in each field in turn for two or more years. This requires careful planning for both time and place and the right division of the farm into smaller units. But once correctly established, a beneficial rotation can continue indefinitely.

Q. Is crop rotation necessary when organic farming methods are used?

A. Many organic farmers claim they can grow corn, wheat and other crops continuously (that is, without rotating) because they practice organic farming so closely they are able to maintain the organic matter content of their soil at a high level. However it is still considered the best practice to employ a good rotation schedule, coupled with use of other organic principles.

Aeration

III

Q. How important is soil aeration?

A. Air is needed in the soil for the proper workings of bacteria and fungi. It aids in the breakdown of organic matter. This is important to the decomposition of the roots of the previous crop. With sufficient air these roots will turn to humus in time to feed the next crop. Air also aids in the oxidation of mineral matter. In an air-poor soil not much of the minerals will be available for plant sustenance. The presence of sufficient air acts as a regulator of the supply of carbon dioxide, too much of which is detrimental to plants. In the process of soil respiration, oxygen is fed to the roots. Better aeration provides a bigger root system and higher yields.

Q. How can soil aeration be promoted?

A. Among the methods used for increasing the air supply in soils are the addition of organic matter, the application of rock powders, soil drainage, subsoiling, cultivation and mixed cropping. By far the most important is keeping the soil supplied with sufficient organic matter. It is axiomatic in agricultural literature that the more humus present in the soil, the better the aeration and the more pore spaces it will contain.

pH (Acid or Alkaline?)

IV

Q. What is pH and how does it affect plants?

A. Soil pH is a way of expressing the amount of acidity or alkalinity. The pH scale runs from 0 to 14. The zero end of the scale is the acid end, while the 14 is on the alkalinity extreme. The neutral point is 7.0; lower than that is acid and higher is alkaline. In general, most common vegetables, field crops, fruits and flowers do best on soils that have a pH of 6.5 to 7.0—that is, slightly acid to neutral. If soils have plenty of humus, most plants will do well even if the soil has a lower or higher pH. A few plants such as azaleas, camellias and gardenias do best on a quite acid soil. If soil is too acid, crushed limestone should be applied. Wood ashes, marl or ground oyster shells are also effective. All of these can be added in compost.

Soils which are too alkaline may be brought back to a favorable pH range by the addition of organic matter. Organic matter contains acid-forming material and produces acids directly on decomposition. These acids combine with any excess alkali, thus neutralizing it.

Q. How can the acid or alkaline condition of a soil be determined?

A. In addition to the value and information gained through a complete soil test, a great deal can be learned concerning the fitness of soil for growing plants by observing what weeds grow in it. An abundant growth of sheep sorrel, Rumex acetocella,

SOIL

Take top three inches of soil when testing.

Mix soil with equal amount of water for litmus test.

indicates that the soil is extremely acid; nettles, violets and various sedges indicate the soil is more or less neutral; while rough bedstraw, Galium aparine, clover, sweet clover, fall dandelion, plantain and Plantago media indicate an alkaline soil. Soil reaction can also be easily tested with litmus paper or by using any of several available indicators.

Moss on the soil indicates either poor drainage, acidity or deficiency of certain elements. A good rule to follow is that where weeds grow, vegetables and other crop plants will also grow.

Q. Can flowers be grouped by their needs for acidity in the soil?

A. Cultivated flowers fall into two groups based upon soil preference. One group will grow only in acid soil with a pH below 6.5, while the others prefer or will tolerate only alkaline soil, pH 6.5 or above.

Acid-lovers are plants that thrive on raw humus, such as their ancestors found in the woods where leaves drop from the taller trees. Leaves, leafmold, peat moss, or other humus should be incorporated in soil where they are to be planted. Especially recommended are oak leaves, which produce an acid humus. The addition of bone meal supplies phosphorus; cottonseed or blood meal supplies nitrogen. These materials may be obtained from the larger seed houses. Dry chicken manure, sheep manure, or compost can be added to leaf mulches as one-fourth of the total.

Q. Will flowering plants which prefer alkaline soil grow in the same bed as those which grow only in acid soil?

A. There are acid-loving evergreens that need very little nitrogen, no rich compost, just leaf mulches and occasionally some bone meal. Likewise there are lilies that like humus and a deep mellow shaded soil; these two can go together. But will a lime-loving plant, like the campanulas grow next to them? Theoretically this looks like a real puzzler. The inexperienced gardener thinks that he should prepare different subsoil and make for himself a great deal of extra work. Practically, the matter is not serious at all. Many hundreds of thousands of gardens grow both azaleas and columbines; both campanulas, foxgloves, coreopsis, carnations and rhododendrons. It seems, indeed, as if the question of acidity were over-emphasized. Organic matter in the ground will act as a buffer against extremes and enable the plants to seek with their roots what they need, lime as well as acid humus. But since the acid-loving shrubby plants will usually be in the background, more leaves may be left there, while in the foreground, where carnations and tulips may be planted, a slight sprinkling of lime is made in fall.

Microorganisms

V

Q. What are the soil microorganisms?
A. There are many kinds and weights of microorganisms in a surface foot of soil and there are large numbers of each kind. Each kind of organism plays some significant role in the decomposition of plant and animal residues, liberation of plant nutrients, or in the development of soil structure. Many groups are dependent on each other: consequently, one kind may tend to follow another.
Q. What are the basic groups of soil microorganisms?
A. Phages and viruses: These are the smallest forms of living matter in the soil. The phages cause diseases of bacteria and the viruses cause diseases of higher plants.

Bacteria: these are the largest group of micro-organisms in the soil. There are many different types. In the process of obtaining their food from plant and animal residues, bacteria in soil bring about the decomposition of these materials.

Some of the important soil bacteria are the ones that convert unavailable nitrogen of the soil organic matter to ammonia, and those that convert ammonia to nitrites and then to nitrates. Others are the bacteria in the root nodules of legumes that fix nitrogen. Most of the nitrogen that is returned to the soil from sources outside the soil is fixed by the legume bacteria. Many other bacteria play important roles in the soil. They make nutrients available or unavailable, modify structure and change air relations of the soil.

Actinomycetes: The characteristic odor that is evident in newly plowed soil in the spring is due to substances produced by actinomycetes. Some of the organisms belonging to this group produce plant diseases such as the potato scab. Many carry on the essential activities of decomposing organic matter and making mineral nutrients available for higher plants. A good soil may have from 100 to 1,000 million bacteria in a gram of soil. Five percent of these are generally actinomycetes.

Yeasts: These single-celled organisms are like bacteria, except they are larger and their structure is more highly developed. The yeasts make up only a small percent of the total organisms in the soil. The importance of yeasts in the soil is not known.

Fungi: The fungi are an essential part of the soil's microbial flora. Although the fungi may be out-numbered by the bacteria, they have a greater mass of growth. Fungi grow best in an aerated soil. Many of them cause plant diseases. However, they decompose organic matter and during the decomposi-

21

tion of plant and animal residues, they synthesize some organic matter as cell tissue.

Algae: These are microscopic plants that form chlorophyll in the presence of sunlight. They are found in the surface layer of soil that is moist and, where light is available, grow as green plants. In the absence of light, they grow as other soil microorganisms. The development of algae may result in the soil turning green at the surface in moist, shady areas. This is not injurious to plants.

Protozoa: These organisms are the simplest form belonging to the animal group. Although they are unicellular and microscopic in size, they are larger than most bacteria and more complex in their activities. Soil may contain as many as 1,000,000 per gram. Protozoa obtain their food from organic matter in the same way as bacteria.

Q. What factors in the soil environment affect the microorganism population?

A. Factors of considerable importance are temperature, moisture, aeration and acidity or alkalinity.

The optimum temperature for a high state of activity is about 85 to 90 degrees Fahrenheit. In order for microorganisms to decay plant material and develop nitrates at a rapid rate, the soil must be warm.

Moisture influences the decomposition of plant and animal residues. When the soil is too dry, there is little or no microbial activity. When the soil has optimum moisture, the beneficial groups of microorganisms are most active.

Generally, a well-ventilated soil supports the growth of beneficial microorganisms that convert nutrients to available forms essential for high crop productivity. Soils possessing good structure are usually well aerated.

Certain organisms may become inactive in acid soils. In general, fungi are more active in acid soils than are bacteria. In more alkaline soils, the actino-mycetes become active. Soils that are excessively alkaline may be devoid of the proper kinds of micro-organisms or the activity of the microorganisms may be limited or directed along lines that are un-favorable for plant growth.

Humus

VI

Q. What is humus and how does it differ from other organic matter?

A. Humus is organic matter which is in a more ad-vanced stage of decomposition than is compost in its early stages. In a compost heap some of the organic matter has turned to humus, but the remain-ing fraction will complete the decomposition process after it has been placed in the soil. Until then, it cannot be called humus. It must still be called organic matter.

The process in which organic matter turns into humus is called humification. The great noticeable difference between humus and organic matter is that the latter is rough-looking material, such as coarse plant matter, while in the humus form we find something that has turned into a more uniform substance.

Q. What are the characteristics of humus?

A. Humus in scientific terminology is called an amor-phous substance, or something with no determinate shape. Humus takes in a vast mixture of compounds, most of which are unknown in formula, broken down to a fine state. It contains both organic and inorganic compounds. It is humus which imparts rich dark color to soil.

Q. How is humus formed?

A. The transformation of organic matter to humus results from microbial and chemical action. In this process, simple compounds are formed which can function directly or indirectly as nutrients. Before its decomposition, organic matter cannot serve as plant food. Only after the humus becomes oxidized to carbonic acid, water, nitric acid and other substances will it work to nourish vegetation.

Q. Of what value is humus?

A. When you apply fresh organic matter to the land, provided conditions are right, decomposition begins almost as soon as the organic matter is covered with earth. (If it is left on the soil surface, the rate of decay will be much slower.) When some of the organic matter turns to humus the rate of decay slows down. The humus, itself, breaks apart slowly as it gives some of its nutrients to the soil solution.

Erosion

VII

Q. Is there a particular cover crop which will keep soil from eroding on steep slopes?

A. Kudzu: This is a warm season crop. Try planting a small area first, maybe ¼ acre, to find out how to handle it. In some areas Kudzu spreads rapidly, even to overtaking trees. It is most difficult to check. It may be destroyed by plowing under when growth starts in the spring. Get all possible information on this before planting.

Q. How did the United States' government counteract the disastrous Dust Bowl soil erosion in the mid-1930s?

A. U.S. Soil Conservation Service specialists, along with government WPA laborers and Dust Bowl area residents, worked on an emergency basis to halt the worst soil-blowing and to level shifting, dune-scarred land. They used snow fences and piles of filled burlap bags, quickly sowing cover crops at the same time. Many of these plantings blew out or dried before germinating; as many as three seasons' plantings were needed before the crops became established.

Long-range measures were begun at the same time and they still serve as exemplary methods for repair or prevention of serious soil erosion. Among these measures are: planting trees as windbreaks and constructing irrigation and stockwater ponds; strip-cropping (plowing only alternate rows for planting); contour plowing (terracing of slopes); "listing" (emergency roughing of strips of dry soil to inhibit blowing); "sweep" plowing (loosening soil from beneath rather than turning it over); and "stubble mulching" (growing heavy-stalk crops such as grain sorghums and leaving the residue as cover between plowings).

Soil Builders: TOP ROW: (l. to r.) wood ashes, basic slag, leaves, sawdust, blood meal. SECOND ROW: raw phosphate, greensand, bone meal, peat most, cottonseed meal, THIRD ROW: seaweed, dried manure, tankage, peanut shells, leaf mold. FOURTH ROW: wood chips, manure, cocoa bean shells, compost, grass clippings.

CHAPTER 3

FERTILIZERS

When the nature of the soil has been determined, added fertilizer is likely to be one of the requirements for insuring the best yield. Why choose organic fertilizers over the chemical ones that many of your neighbors use? One basic reason is the devastating effect concentrated chemicals can have on the soil. Another is the versatility of natural fertilizers. They contain a wide variety of nutrients which are gradually, methodically allotted to the soil. You can get essential elements into the soil in meaningful amounts without resorting to bags of chemical compounds.

You should know that manure, valuable as it is for some crops, can damage others or reduce the quality of the harvest. Pay attention to the kind of plants you have and the type of manure you have access to.

If you want to add minerals to the soil, ground rock fertilizers are difficult to beat. Follow the general rules about how much should be applied to specific types of plants according to the space involved.

Sewage waste is rich in many plant nutrients and it makes an excellent fertilizer. But it must be treated to destroy the dangerous germs and control the spread of disease.

Chemical Vs. Organic

I

Q. What are the different types of fertilizers and how does one differentiate among them?

FERTILIZERS

A. a) A natural fertilizer is one that consists of some natural earth product which may be processed mechanically, but is not treated with acids or other substances to increase its solubility. Phosphate rock, finely pulverized, is a natural fertilizer.

b) An artificial, chemical fertilizer is a combination of some earth product and a strong acid. Superphosphate, which is made by treating phosphate rock with sulfuric acid, is an example. A hundred pound bag of superphosphate includes 50 pounds of phosphate rock and 50 pounds of sulfuric acid.

c) An organic fertilizer is a mix of plant and animal residues. These may be fresh residues, or residues which have accumulated and have been preserved for long periods of time, such as peat, marl and limestone.

d) A raw, organic fertilizer is made up of raw (unfermented) plant and animal residues. Or it may be made of raw, organic matter to which such materials as pulverized phosphate rock, potash rock, oyster shell flour and seaweed have been added; these contain most of the elements which are apt to have been leached out of the soil.

Q. Why do organic gardeners disapprove of artificial, chemical fertilizers?

A. Chemical fertilizers are quick-acting, short-term plant "boosters" and are responsible for: (1) deterioration of soil friability, creating hardpan soil, (2) destruction of beneficial soil life, including earthworms, (3) altering nutritional content of certain crops, (4) making certain crops more vulnerable to diseases, and (5) preventing plants from absorbing some needed minerals.

The soil must be regarded as a living organism. An acid fertilizer, because of its acids, dissolves the cementing material in the soil. This material is made

28

up of the dead bodies of soil organisms and holds the rock particles together in the form of soil crumbs. Acid fertilizers spoil the friability of the soil. On the surface of the soil, such cement-free particles settle to form a compact, more or less impervious layer. This compact surface layer of rock particles encourages rain water to run off rather than enter the soil.

Q. Do chemical fertilizers interfere with soil aeration?

A. There are several ways by which artificial fertilizers reduce aeration of soils. Earthworms, whose numerous borings make the soil more porous, are killed. The acid fertilizers will also destroy the cementing material which binds rock particles together in crumbs. Lastly, hardpans result which seal off the lower soil levels, keeping them more or less completely anaerobic.

Q. How is hardpan formed by using chemical fertilizer?

A. A highly soluble fertilizer, such as 5-10-5, goes into solution in the soil water rapidly, so that much of it may be leached away without benefiting the plants at all. But the sodium in the fertilizer, like sodium nitrate, tends to accumulate in the soil where it combines with carbonic acid to form washing soda, sodium carbonate. This chemical causes the soil to assume a cement-like hardness. Other minerals, when present in large concentrations, percolate into the subsoil where they interact with the clay to form impervious layers of precipitates called hardpans.

Hardpans seal the topsoil off from the subsoil. Water cannot pass downward into the subsoil, and water from the table cannot rise to the topsoil where the plants are growing. Many plants cannot live when their roots are too wet. Then too, the subsoil below the hardpans is anaerobic and rapidly becomes acid. In such anaerobic, acid soils, the soil organism changes radically and in ways which are unfavorable to crop plants.

Q. What damage do the chlorides and sulfates do?

A. Such highly soluble chemicals as chlorides and sulfates are poisonous to the beneficial soil organisms, but in small amounts they act as stimulants. These chemicals stimulate the beneficial soil bacteria to such increased growth and reproduction that they use up the organic matter in soil as food faster than it can be returned by present agricultural practices. When chemical residues accumulate in the soil, the microorganisms may be killed off by hydrolysis (water-removing). The high salt concentration in the soil water will pull water from the bacterial or fungal cells, causing them to collapse and die.

Many artificial fertilizers contain acids, such as sulfuric and hydrochloric, which will increase the acidity of the soil. Changes in the soil acidity (pH) are accompanied by changes in the kinds of organisms which can live in the soil. Such changes are often sufficient to interfere greatly with the profitable growth of crop plants. For this reason, the artificial fertilizer people tell their customers to use lime and to increase the organic matter content of their soil, thus offsetting the deleterious effects of these acids.

Q. How does the use of chemical fertilizer make plants more susceptible to diseases?

A. Chemical fertilizers rob plants of some natural immunity by killing off the "policemen" microorganisms in the soil. Many plant diseases have already been considerably checked when antibiotic-producing bacteria or fungi thrive around the roots. When plants are supplied with much nitrogen and only a medium amount of phosphate, plants will more easily contract mosaic infections. Most resistance is obtained if there is a small amount of nitrogen and a large supply of phosphate. Fungus and bacterial diseases have been related to high nitrogen fertilization and to lack of trace elements.

Probably the most regularly observed deficiency in plants doped continually with artificial fertilizers is in trace minerals. To explain this principle will mean delving into a little physics and chemistry, but you will then easily see the unbalanced nutrition created in artificially fertilized plants. The colloidal humus particles are the vehicles that transfer most of the minerals from the soil solution to the root hairs. Each humus particle is negatively charged and will, of course, attract the positive elements, such as potassium, sodium, calcium, magnesium, manganese, aluminum, boron, iron, copper and other metals. When sodium nitrate, for instance, is dumped into the soil year after year in large doses, a radical change takes place in the humus particles. The numerous sodium ions (atomic particles) will eventually crowd out the other ions, making them practically unavailable for plant use. The humus becomes coated with sodium, glutting the root hairs with the excess. Finally, the plant is unable to pick up the minerals that it really needs.

Q. How can I determine the quality of my soil and find out what is needed to enrich it?

A. There are two basic soil-testing methods accessible to the average gardener or farmer. The simplest is the home soil-test kit, which is marketed commercially and has been used successfully by many thousands. The kit provides solutions and guides for determining the approximate content of the three major soil nutrients—nitrogen (N), phosphorus (P) and potash (K)—plus a similar means of establishing the pH standing. Home testing kits are easy to use, require no knowledge of chemistry or laboratory procedure. One of their biggest advantages is that they permit you to make frequent on-the-spot tests of your soil (whereas other methods are often delayed) and help you adapt a fertilizing program in accord with your soil's prime needs.

The other general way to have soil tested is to send a sample to a laboratory offering this service or to your state experiment station. Farmers, rural homesteaders and city gardeners frequently get help and information through their county agent. Most state experiment stations charge a small fee for the soil test report, which in addition to an NPK and pH analysis may include indications of trace element levels and organic matter content.

Nitrogen, Phosphorus, Potassium
II

Q. What are the best sources of nitrogen?

A. To add nitrogen to soil, use tankage, manure (fresh or dried), homemade or commercial compost, sludge, or any of the vegetable meals such as cottonseed, linseed, soybean or peanut. Because of the high nitrogen content of blood meal and dried blood (12 to 15 per cent), use these materials more sparingly.

For vegetable gardens and pastures cottonseed meal is one excellent source of nitrogen. When you consider what you are getting in the meal compared to nitrate of soda which costs only a little less, you are definitely better off using the meal. General early spring application rates are 200-300 pounds per acre or heavier depending upon conditions, five pounds per 100 square feet for flower and vegetable gardens, and the same rate for lawns.

Blood meal, costing about the same as cottonseed meal, is quicker-acting because its nitrogen is readily available. Fish meal is another good N source, especially for corn (about two pounds per 60-foot row). Generally speaking, use organic fertilizers which are easily obtainable in your area.

Q. How much phosphate should be applied per acre of land?

A. When we recommend phosphate to a farmer, we want him to use 1500 to 2000 pounds per acre on his first application. Following this up on each rotation with 1000 pounds plus crop residue will increase the mineral content as well as the fertility of soil. Many farmers go by the rule: "Feed your phosphate to your clover, feed your clover to your corn, and you won't go wrong."

The trouble of "too light applications" often pops up with rock phosphate. Even after a soil test shows a phosphorus deficiency some new organic growers continue to apply less than 500 pounds of rock phosphate per acre. It is impossible to build up the phosphate level with this small an application.

One thing a farmer or gardener must remember: to get best results, you must have your phosphate level built up along with your organic matter. The supply of phosphate in our virgin soils was about 1,200 pounds per acre. In many areas, this level is now down to 125 pounds per acre.

Q. How can potash be restored to depleted land?

A. Green manuring with deep-rooted crops like sweet clover and alfalfa, helps to restore potash to depleted land. Greensand and granite dust applied at from 100 to 1,000 pounds per acre will definitely help correct the problem. Many farmers feed phosphate to small grains and greensand to their corn. Many we know use 500 to 1,000 pounds per acre through the corn drill when they plant and are harvesting 100 bushels per acre.

Q. Does this affect their nutritional value?

A. Plants grown with artificial fertilizers tend to have less nutrient value than organically grown plants. For example, several experiment stations have found that supplying citrus fruits with a large amount of highly soluble nitrogen will lower the vitamin C content of oranges. These fertilizers that provide quickly soluble nitrogen will also lower the capacity of hybrid corn to produce seeds with high protein content.

Q. What is the importance of nitrogen in plant production?

A. Nitrogen is responsible for producing leaf growth and greener leaves; lengthening the growth period tends to increase yield of fruits. Deficiency in nitrogen causes yellow leaves and stunted growths; excess delays flowering, causes too much elongation of stem, reduces quality of fruits, causes lodging of wheat and renders crops less resistant to disease.

Q. Why is nitrogen from an organic source better than nitrogen from a chemical fertilizer?

A. Organic forms of nitrogen are more stable in the soil. They become available for plant growth more gradually than nitrogen from artificial fertilizers. When concentrated chemical nitrogen is applied to the soil, it produces a shot-in-the-arm effect to plant growth. The plants are subjected to too much nitrogen at one time. Then, if a sudden heavy rain-storm drenches the field, the chemical form of nitrogen is, to a large extent, washed out, and the plants can be starved for lack of the element.

Q. What is the significance of phosphorus in agriculture?

A. All growing plants need phosphorus. It is important for a strong root system, for brighter, more beautiful flowers and for good growth. If plants are unusually small or thin, with purplish foliage, it may be an indication of a phosphorus deficiency in the soil. Phosphorus is also said to hasten maturity; it increases seed yield, fruit development, resistance to winterkill and disease, and bolsters vitamin content of plants. A phosphorus deficiency causes stunted growth and sterile seed.

Q. What are the best sources of phosphorus for the organic gardener?

A. Phosphorus can best be added to soil through the application of rock phosphate, a natural rock product containing 30 to 50 per cent phosphorus. When the rock is finely ground, the phosphate is available to the plant on demand, as it were. Rock phosphate is especially effective in soils which have organic matter. Besides rock phosphate, other phosphorus sources are basic slag, bone meal, dried blood, cottonseed meal, and activated sludge.

Q. What is the role of potassium in plant nutrition?

A. Potassium, the third major nutrient, strengthens the plant. It carries carbohydrates through the plant system, helps form strong stems and helps the plant to fight diseases. If the plants are slow-growing and stunted, with browning leaves and under-sized fruits, there is probably a potassium deficiency.

Potassium is effective in: improving the keeping quality of fruit; aiding in the production of starches, sugars and oils; decreasing the water requirement of a plant; making plants more disease resistant; reducing winterkill and deepening the color of fruit. Potassium is essential for cell division and growth; it helps plants to use nitrogen, balances the effects of excess nitrogen or calcium and reduces boron requirements. Deficiency of this mineral can cause firing of the edges of leaves—lower leaves first—resulting in shriveled, sterile leaves. Corn ears frequently fail to mature properly when there is a potassium deficiency.

Q. What is the fertilizing value of fish scraps and how should they be used?

A. Fish scraps contain about seven per cent nitrogen and about the same amount of phosphorus. Fish scraps abound in trace or minor elements, which is why fish is an excellent organic fertilizing material. Because fish scraps are rich in fish oil they may attract fat-eating ants. Dried, ground fish is fat-free and breaks down faster.

When using fish scraps in compost heaps, cover the pile with a thin layer of soil. This will reduce the odor of the fish material. Be sure the pile is well moistened to assist in rapid decomposition.

Manure

III

Q. What crops benefit most from the application of manure?

A. Grasslands are generally much benefited by top dressing with farm manure, either fresh or fermented.

Root crops usually respond satisfactorily to generous applications of stable manure. Some precautions, however, must be taken in the case of potatoes and sugar beets. Excessive amounts of fresh manure on light soils and loams cause the beets to become very large but with a low sugar content. Fresh manure can result in leggy plants.

Corn, millet and leafy crops in general respond favorably to manure, fresh or decomposed.

Garden crops respond quite favorably to applications of manure. For every 100 square feet of garden, approximately 25 pounds of manure mixed with 10 pounds of rock phosphate is enough to enrich and condition soils. If the manure is not well-rotted and larger quantities are used, apply it three or more weeks before planting and incorporate the manure into the soil.

Young, deciduous trees and shrubs respond favorably to manure, but prefer the decomposed.

Q. What's the best way to apply manure?

A. Many authorities recommend getting the raw manure into the ground immediately, plowing it under before any nutrients escape. Speed is all-important; therefore, ideally, the manure spreader should make its rounds every day.

Q. How should manure be stored?

A. As everyone knows, it isn't always possible to spread manure daily because of the weather, time of year or other work. In such cases manure must be stored. And when it comes to storing manure, a cement floor pays for itself in one year by just about ending leaching.

The pile should be about four feet high, steep-sided and flat-topped. No cover is needed since the manure pile usually receives enough rain to prevent heating, but not enough to cause much leaching. All authorities agree that heating or fermentation are to be avoided in order to conserve nitrogen and other nutrients.

Storage in loose-run barns or sheds with concrete floors is also recommended—if plenty of bedding or litter is used. In this case, the manure is permitted to accumulate while the urine keeps it moist and the animals trample it compact.

Q. How does manure lose its nutrients?

A. The two chief manure robbers are *leaching* and *evaporation*. Together these two forces steal considerable amounts of the active nutrients in manure —one by washing them out, the other by drying them off into the air.

Q. Does it make any difference what kind of manure is used?

A. Yes, different manures have varying amounts of nitrogen, phosphorus and potash. The following chart should help you decide which one best suits your needs.

37

Kinds of Manure	Moisture	Nitrogen	Phosphorus	Potash
Cattle (fresh excrement)......		0.29%	0.17%	0.10%
Cattle (fresh urine)..........		0.58	....	0.49
Hen manure (fresh)..........		1.63	1.54	0.85
Horse (solid fresh excrement)..		0.44	0.17	0.35
Horse (fresh urine)..........		1.55	....	1.50
Human excrement (solid).....	77.20	1.00	1.09	0.25
Human urine...............	95.90	0.60	0.17	0.20
Night soil.................		0.80	1.40	0.30
Sheep (solid fresh excrement)..		0.55	0.31	0.15
Sheep (fresh urine)..........		1.95	0.01	2.26
Stable manure, mixed........	73.27	0.50	0.30	0.60
Swine (solid fresh excrement)..		0.60	0.41	0.13
Swine (fresh urine)..........		0.43	0.07	0.83

Q. How does fresh manure differ from rotted manure?

A. If we assume that fresh manure is a normal mixture of urine and feces and that conditions have been controlled, fresh manure differs from rotted manure in composition as follows:

1. Rotted manure is richer in plant nutrients. This is a result largely of the loss in dry weight of the manure. One ton of fresh manure may lose one-half its weight in the rotting process.

2. The nitrogen in the composted (rotted) manure has been fixed by microorganisms. These organisms build the soluble nitrogen compounds into their own bodies. The nitrogen in the urine is used in the formation of complex proteins during the decomposition of the manure.

3. The solubility of the phosphorus is greater in the composted manure. If leaching can be prevented, there is no change in the total amount of phosphorus and potassium. Precautions must be taken to prevent the loss of nitrogen in the composting process.

Rock Fertilizers

IV

Q. Why are ground rock fertilizers the best sources of minerals for the soil?

A. Organic matter acting on crumbling rock dust is the way soil is generally formed. The rock basis for soil produces a continuing supply of mineral-rich silt that keeps breaking down, and in large part replaces the minerals that crops take from the soil.

Powdered rocks contain a large number of mineral elements which are absolutely essential for normal and healthy plant growth. They provide two major nutrients, phosphorus and potassium, in a long-lasting, safe and non-caustic form. In addition, they are a rich source of trace elements, especially necessary to keep plants disease-free and pest-free. Powdered natural rocks contain the elements which have been leached out of our agricultural soils during the past 50 to 100 years of modern, not necessarily intelligent, farming and gardening. Natural rock fertilizers have the following advantages: they provide a "balanced diet" for the soil; they supply long-lasting nutrients that eliminate the need of frequent applications; they improve the soil's physical structure and water-holding capacity; and they do not leave harmful after-effects that slow down the activities of soil bacteria.

It is important to remember that one application of natural rock fertilizers will last as long as 5 to 10 years. When you use natural rocks, you are, in a way, putting plant nutrients in the soil bank instead of feeding the plant hand-to-mouth.

Q. Are there any general rules about how much of the rock fertilizers should be applied to different plants?

A. Yes. Here are the basic ones:
   a) Gardens—(all vegetables) 10 pounds per 100 square feet or apply directly in hills or rows, with plants and seeds;
   b) Roses and bushes—1 pound each;
   c) Flower beds—5 to 10 pounds per 100 square feet;
   d) Lawns—5 pounds per 100 square feet;
   e) Trees—(all kinds) 15 to 100 pounds per tree. Spread to edge of drip line;
   f) House plants—1 tablespoon to 5-inch pot, 1 teaspoon to smaller pots;
   g) Corn, soybeans, oats, wheat and rye—250 to 400 pounds in drill or planter attachment, or an application of 500 to 1,000 pounds broadcast per acre;
   h) Alfalfa and all clover crops—new seeding or at time of planting, 500 to 1,000 pounds per acre.
   A thorough soil test is the best guide for the needs of individual gardens or farms and their plants. Mineral requirements supplied by other natural fertilizers used should also be considered.

Trace Elements

V

Q. What are the trace elements and how important are they?
A. Trace elements are minor mineral nutrients needed by all plants, animals and humans in extremely small, or "trace" amounts. These micronutrients must be available in the soil in which the plants and foods are grown. Shortages of one or more of these can result in plant or animal disease. On the other hand, too much of any trace element opens the door to a host of toxic conditions in plants and sicknesses in animals and people.

Trace elements constitute less than one per cent of the total dry matter of the plant, but don't minimize their importance. They sometimes determine the vigor of the plant. Even where good crops have been produced, trace elements added to the soil have raised yields and improved crop quality.

Q. If soils lack a particular trace mineral element, such as boron, manganese, copper, or cobalt, is it advisable to buy one of these elements to correct the deficiency?

A. No. That is a dangerous procedure. The amounts needed are usually small, as little as five parts per million. Ten parts per million could damage crops.

Q. How does an organic gardener supply these trace elements?

A. The safest, most reliable method for assuring an adequate supply of the minor elements is through the use of organic fertilizers and the use of certain plants which accumulate these minerals. Compost, leaf mold, mulch, natural ground rock fertilizers and lime help provide a complete, balanced ration of both major and minor nutrients.

The soil, like human beings, should have a varied diet. Some other good sources of trace elements are seaweed and fish fertilizers, weeds that bring minerals up from deep in the sub-soil, garbage compost, and sewage sludge which contains wastes from all over the world. Besides supplying trace elements themselves, these materials, when decomposing, make other elements already present in the soil available.

Taboos

VI

Q. How do the highly soluble artificial fertilizers harm soil organisms?

FERTILIZERS

A. Highly soluble chemicals such as chlorides and sulphates are poisonous to beneficial soil organisms, but in small amounts act as stimulants. These chemicals stimulate the growth and reproduction of the beneficial soil bacteria so much that the bacteria use up the organic matter in the soil as food faster than it can be returned by present agricultural practices. When chemical residues accumulate in the soil, the microorganisms may be killed off by hydrolysis (water removing). The high salt concentration in the soil water will pull water from the bacterial or fungal cells, causing them to collapse and die. Earthworms will be poisoned by ingesting soil and humus particles coated with chemical residues.

Q. What about combining organic methods with chemical fertilizing?

A. Even though the incorporation of organic methods into chemical gardening and farming will greatly improve general results, 100 per cent organic methods are far superior. Additions of chemicals can prove toxic to soil life and plants. They tend to produce an "off flavor" in vegetables and can never achieve the perfect balance of soil nutrition which is inherent in natural fertilizers.

Q. It has been reported that servicemen overseas are getting hepatitis unless they have gamma globulin shots to protect them, because the sewage from homes is used as fertilizer for food crops. Is fertilizer for gardens made from sludge?

A. Sewage waste, being rich in nitrogen and other plant nutrients, makes excellent fertilizer, *but not if it is untreated.* Raw sewage may contain various disease organisms. In this country sludge is dehydrated and processed so that no pathogens survive. Answering this question in a syndicated newspaper health column, Dr. Joseph G. Molner explained that ". . . the processing amounts to a particularly effective form of pasteurization. The germs are destroyed and the fertilizer becomes safe to use. In many countries, there is no processing and the germs remain."

Miscellaneous

VII

Q. What are the correct amounts of fertilizers for very small gardens based on the recommendations usually given in much larger quantities?

A. Using these figures, you can figure fertilizer applications for small areas from given rates per acre. For each 100 pounds per acre, equal rates for small areas would be as follows:

> for 1,000 square feet—2½ pounds or 2½ pints
> 100 square feet— ¼ pound or ½ cup
> 1 square yard—½ ounce or 2½ teaspoonfuls

For each 2,000 pounds or one ton per acre, equal rates for small areas would be:

> for 1,000 square feet—50 pounds
> 100 square feet— 5 pounds or 5 pints
> 1 square yard—½ pound or 1 cup

Amount of Fertilizer Per Row for Various Rates of Application

Approximate amounts per 100 feet, for rows different distances apart

Rate per acre	12 in.	15 in.	18 in.	24 in.	30 in.	36 in.
250 pounds...	9 oz.	12 oz.	14 oz.	1 lb.	1¼ lbs.	1½ lbs.
500 pounds...	1 lb.	1¼ lbs.	1½ lbs.	2 lbs.	2½ lbs.	3½ lbs.
750 pounds...	1½ lbs.	2 lbs.	2½ lbs.	3 lbs.	3¾ lbs.	4½ lbs.
1000 pounds...	2¼ lbs.	2½ lbs.	3 lbs.	4½ lbs.	5¾ lbs.	7 lbs.
1500 pounds...	3½ lbs.	4 lbs.	5 lbs.	6½ lbs.	8½ lbs.	10½ lbs.
2000 pounds...	4½ lbs.	5 lbs.	6½ lbs.	9 lbs.	11 lbs.	13½ lbs.

The amounts of fertilizer required on small areas or for row feet may be calculated from this table.

—Pennsylvania Agricultural Extension Service

Q. What is greensand and how is it beneficial?

A. Greensand is a naturally occurring marine deposit of glauconite, a mineral-rich iron-potassium silicate.

Also called greensand marl, it contains six to seven per cent potash, actually more than granite dust which contains from three to five per cent. Being an underseas deposit, it contains most of the elements found in the ocean and is an excellent soil-builder. Greensand has the ability to absorb large amounts of water and provides an abundant source of plant-available potash. Superior deposits contain 18 to 23 per cent iron oxide, 3 to 7½ per cent magnesium, and small amounts of lime and phosphorous, as well as potash.

Greensand is versatile. It may be applied directly to the soil and plant roots, left on the surface as a combined mulch-compost, or used in compost heaps to stimulate bacterial action and enrich the heap.

CHAPTER 4

COMPOST

Composting is central to the idea of organic gardening, for it is actually an imitation of nature's way to rebuild the soil by encouraging decomposition of natural plant particles. Composting provides the gardener with a means of turning waste into the kind of rich, dark humus he would ordinarily buy, yet it costs nothing to make compost and it requires very little effort. There are several effective methods of composting, so you have a choice as to which you find simplest and most convenient.

The materials you can use are almost unlimited in variety—anything from table scraps to weeds to fallen autumn leaves.

If you are planning to make compost for the first time, you will find the editors' directions for building simple and inexpensive bins useful and surprisingly easy. Compost is a great help in flower or vegetable gardens, in raising trees or transplanting shrubs, in any area of indoor or outdoor planting.

Importance of Composting

I

Q. How can compost be defined?
A. The compost heap in your garden is an intensified version of the process of death and rebuilding which is going on almost everywhere in nature. In the course of running a garden, there is always an accumulation of organic waste of different sorts

Making compost by the Indore Method.

—leaves, grass clippings, weeds, twigs—and since time immemorial gardeners have been accumulating this material in piles, eventually to spread it back on the soil as rich, dark humus.

Because the compost heap is symbolic of nature's best effort to build soil and because compost is the most efficient and practical fertilizer, it has become the heart of the organic method. It is the basic tool necessary to do the job—creating the finest soil. The material *must be decomposed* to be called compost, and it is not to be called compost until it is. There are two elements necessary for a compost heap: organic materials and the proper conditions to make them decompose. The degree of decomposition is referred to as finished or unfinished. In a more or less finished compost, the materials are greatly reduced in the extent of their fibrous appearance. Organic matter is the raw material of the composting process; its bacteria hastens the formation of humus. Compost is more than a fertilizer or a healing agent for the soil's wounds—it is a symbol of continuing life.

Q. What are the most common methods of composting?

The Indore Method:

A. The first organized plan for composting was perfected by Sir. Albert Howard. To utilize the Indore Method, place a layer of brush on the ground to provide a base.

Then build the heap in layers, using first a six-inch layer of "green matter" like weeds, crop wastes or leaves. Next comes a two-inch layer of manure, which is in turn covered by a sprinkling of topsoil and limestone. The layers are repeated until the pile reaches a height of about five feet.

The pile is turned after six weeks and again after twelve weeks to allow the air to penetrate all parts of the heap. After three months the compost will be "finished."

The 14-Day Method:

The keystone of the 14-day method is the grinding or shredding of all materials going into the compost pile. Grinding has these effects on compost: (a) The surface area of material on which microorganisms can multiply is greatly increased. (b) Aeration of the mass is improved because shredded material has less tendency to mat or pack down.

47

14 Day Method.

The Anaerobic Method:

Anaerobic method means composting in an environment free of air. The usual method of anaerobic composting is by using a digester, an enclosed container consisting of several sections. The organic waste is introduced into the top section of the digester where it is inoculated with bacteria.

The anaerobic method overcomes the two disadvantages of aerated composting: oxidation, which

48

destroys much of the organic nitrogen and carbon dioxide; and the waste of some valuable liquids which leach into the ground underneath.

One big difficulty has been finding an efficient and simple way to practice anaerobic composting. The lastest technique is to enclose the compost in a plastic wrapping.

Sheet Composting Method:

Sheet composting is most advantageous as a permanent method of rebuilding garden soil. Sheet composting means putting manure and other raw organic matter into the soil fresh, then letting it decay for a month or two while no crops are growing. Inasmuch as organic matter is most valuable to the soil while it is decaying, it is important to allow as much of the decay process as possible to take place in the soil.

Fresh or raw organic material may be applied as is in sheet composting practice. The undecayed material adds more nitrogen to the soil than when it is almost fully decomposed. The reason is this: in the process of decomposition in a large bin, heap or pit, there is considerable heat generated which vaporizes much of the nitrogen. Also, organic material when applied fresh releases its minerals more slowly than when decayed.

Earthworm Method:

For the earthworm method, the compost can be made in boxes indoors or in pits on the outside. The boxes should not be more than two feet high, but their width and length are optional. The pit also can be of any width or length that is convenient; it can be of any shape—square, oblong, or even round. Because earthworms do not like light, make a wooden top, but not too tight-fitting. Remove the cover when it rains.

Place a variety of materials in the pit and mix thoroughly. The more the materials are mixed, the less they will tend to heat up. See that the mass is well watered and put the earthworms in immediately. The more earthworms, the quicker the composting. A pit ten feet square will easily take 10,000 earthworms—and more! One need not purchase more than 1,000 or 2,000 earthworms because they multiply fantastically. The action of the earthworms will help to produce the finest compost you could ever make.

In the boxes, a typical mixture would be about 70 per cent weeds, leaves, and grass clippings, about 15 per cent manure and about 15 per cent topsoil. These elements can be varied. If no manure is available, substitute parts of your table waste. You can try almost any formula. By this method compost can be made in two months or less.

Earthworm method of composting.

Sheet composting—manure and crop residues.

Q. What can be done to speed up the composting process on leaves?

A. Use a rotary mower to shred leaves, weeds, straw, hay and garden wastes of almost any kind. These materials will turn into compost more rapidly when they are shredded. Merely spread the material on the ground and run over it with the mower or mower attachment. If the material is piled rather high, tilt the mower to lift the cutting blades and position them directly over the pile; then slowly lower the blades into the pile. Do this beside a fence, a large cardboard carton or a building that will act as a backstop to pile the material.

It's still important to add some nitrogen-rich material such as fresh or dried manure, dried blood, compost made previously or a small amount of rich soil. The nitrogen of these materials is an essential food for the decomposing bacteria.

51

Shredding compost for use on a lawn.

Material

II

Q. What materials may be composted?

A. With very few exceptions, any organic material is a potential candidate for the compost heap. Any vegetable or animal matter and even some minerals can be returned to the soil through composting.

Q. Where and how are compost materials best obtained?

A. Look around the house and grounds at home; you can see weeds, grass clippings, kitchen garbage,

Lawn clippings, residues and wastes make good compost materials.

leaves, garden residues and other organic matter. Check at nearby farms and offer to buy spoiled hay, manure and other wastes. Another great source of compost material, often untapped: various nearby industries.

Here's a system for securing compost materials that is used successfully by many gardeners: First, page through the classified section of your telephone directory; make a list of a few promising firms (lumber companies, mills, meat packing houses, quarries, dairies, leather tanneries, city park departments, riding stables and wholesale food companies). Then you are ready to embark on a most rewarding collection trip. You can use the trunk of your car to haul materials back to your garden.

Q. Is manure necessary in making compost?

A. Even if animal manures are not available, compost can be made successfully. Cut or shred the plant materials as finely as possible in order to expose a maximum amount of surface to the organisms of decay. As soon as the heat has subsided, the heap of finely ground plant materials may be inoculated with earthworms especially bred for the compost heap and soils rich in organic matter. These worms will supply the manure and the various animal excretions needed. It also may be well to include in the compost heap such animal residues as bone meal, dried blood, dried meat meal, dried fish and dried manure if available.

Q. Can the leaves of any tree be used for composting? I've heard that oleander leaves are poisonous, for instance, and that eucalyptus won't compost.

A. The leaves of all trees are a valuable source of organic matter and minerals. In the case of the oleander it is the tree's sap which is poisonous. No toxic or otherwise harmful qualities would be imparted to your compost by using the leaves correctly, particularly since these would be decomposed, heated in the composting process, and mixed thoroughly with many other materials. Such leaves as those of the eucalyptus, camphor and walnut should be exposed to the weather for a time before composting so that components they contain which might interfere with organisms of decomposition will be leached out. Shredding helps prepare any leaves better for composting.

Q. Can seaweed and kelp be added to the compost heap?

A. Both are high in potash (about five per cent) and trace elements. Many seaweed users apply it fresh

54

Rockweed—a seaweed often used to enrich a compost heap.

from the sea; others prefer washing first to remove salt. It can be used as a mulch, worked directly into the soil, or placed in the compost heap. Dehydrated forms are available commercially.

Q. What are the best materials to use in making compost for blueberry land?

A. Blueberries require an acid soil for normal growth

55

and development. Compost for blueberries should include up to 20 per cent sawdust and 10 per cent oak leaves (finely ground if possible) with such other plant materials as may be available. Instead of using soil in the compost heap, use acid peat. Omit lime or other alkaline soil amendments from the heap.

Q. Are wood ashes useful in gardening?

A. Yes, because they contain 1.5 per cent phosphorus, 7 percent or more potash. Never allow wood ashes to stand in the rain, as the potash will leach away. They can be mixed with other fertilizing materials, side-dressed around growing plants or used as a mulch. Apply about 5 to 10 pounds per 100 square feet. Avoid contact between freshly spread ashes and germinating seeds or new plant roots by spreading ashes a few inches from plants. Wood ashes are alkaline. Mixed with leaves, they speed decomposition.

Q. Are banana wastes of any special value in composting?

A. Yes. Banana skins and stalks are extremely rich in both phosphoric acid or phosphorous and potash (2.3 to 3.3 per cent in phosphoric acid and from 41 to 50 per cent in potash, on an ash basis.) The nitrogen content is considered relatively high. Table cuttings and kitchen refuse containing banana skins are valuable in composting because these materials are rich in the bacteria which cause quick breakdown and encourage decomposition of the rest of the compost material.

Q. Can I use cotton combings and cotton in the compost heap?

A. Cotton combings and cotton may be used in the compost heap as part of the green matter. In addition to organic matter, they contain 1.32 per cent nitrogen, 0.45 per cent phosphoric acid, 0.36 per cent potash, as well as various trace elements. Cot-

56

ton gin boll wastes are also excellent materials, rich in potash and other nutrients.

Q. Is there any worth in human hair for fertilizing and composting?

A. Hair, like wool and silk, has a high nitrogen content. If the sweepings from a barber shop were regularly applied to a compost heap, a great amount of nitrogen could be saved. Six to seven pounds of hair contain approximately a pound of nitrogen—or as much as found in 100 to 200 pounds of manure. If kept in a well-moistened heap, hair will disintegrate as easily as feathers.

Q. What is meant by green matter in compost?

A. The term green matter refers to any and all plant materials which are used in the compost heap or elsewhere as a composting constituent. Green matter may consist of leaves, weeds, grass clippings, corn stalks, sunflower stalks, hedge trimmings, seaweed, spoiled hay or straw, kitchen wastes, chaff, apple pomace, castor bean pomace, vinery wastes, gin wastes, winery wastes, sawdust, grocery store wastes, brewery wastes, garbage or any other kind of plant material you can lay your hands on. In many cases you will find neighboring land growing in weeds. It is not difficult to get permission to get these weeds for the cutting. By cutting these weeds before they go to seed, you can not only get green matter for composting, but you also protect your own land from being seeded by the weeds.

Q. Are some kinds of green matter only available in certain sections of the country?

A. Available green matter differs in different sections of the country depending upon crops grown and the kinds and nature of processing plants. Great quantities of green matter often are to be had at little or no cost because so few people appreciate the great value and need of organic matter in the soil. Green matter which may only be available in particular areas includes apple pomace (pulp, ground fruit or plant residue) castor pomace, coffee grounds, sawdust, sandpaper dust, pecan and other nut shells, wastes from coffee roasting plants, cocoa wastes, cotton gin wastes, shredded palm tree roots, pea vines, flax wastes, beet wastes, grape residues, rice hulls, bean pods, spent hops and other brewery wastes, tomato wastes, pits of stone fruits.

Q. I have been interested for some time in building a compost heap, but I find suitable green matter difficult to come by. Can you help me?

A. Materials for making a compost heap are not as hard to find as you may think. Recently one of our readers told us that he was hunting materials for composting and began to think it was hopeless until quite by chance he parked his car near a supermarket, and noticed that each day a truck load of plant waste was being discarded. He found upon inquiring that it was possible for him to get this waste material at no cost. The grocery store waste plus the manure he could get from a local riding stable enabled him to build all the compost he could use for his garden.

Q. Is there enough green matter to go around for all those who are interested in making compost?

A. The supply of green matter suitable for making compost is almost inexhaustible. The U.S. Forest Service has estimated that some twenty million tons of sawdust are produced each year by the approximately 40,000 sawmills in the United States. Then there are many millions of tons of peat which can be bacterized and composted. Also there are almost limitless amounts of such forest products as leaves and shredded bark; leaves from city streets; by-products from many industries using plant materials, such as hops from breweries, corncobs from grain mills, cotton wastes from cotton mills, apple pomace from cider mills, cannery wastes, garbage, spoiled hay, wastes from pea vineries, nut shells of many kinds, water hyacinths, seaweed, wastes from grocery stores, and innumerable other plant residues available in various parts of the country. It is surprising how much green matter can be found by one who actually goes looking for compostible materials.

Q. Can Spanish moss which hangs from many of the trees in our southern states be useful for making compost?

A. Yes. Spanish moss is not a real moss, but a seed plant belonging to the pineapple family of flowering plants. Spanish moss is harvested, cleaned and sold as a stuffer to mattress and upholstery people. The cleanings outside moss gins accumulate like the sawdust around a sawmill. There are hundreds of moss gins in Louisiana alone, around which are more than a million pounds of moss residues. Composted moss gin wastes are now being sold under the name of "Gro-Mulch."

Q. Are coffee grounds considered fit material for the compost heap?

A. Yes. Coffee grounds are organic matter which have up to two per cent nitrogen, small amounts of phosphorus and varying amounts of potash.

Compost Heaps and Containers

III

Q. Which is best—to make a compost heap gradually as the material collects, or to collect all the material first and then build the heap in one operation?

A. In starting a compost heap it is important to have enough material to get the proper height so that heating takes place. Small additions can then be made to the heap, building on lengthwise. Keep in

Fertilizer can be sprinkled into compost heap.

mind, however, that collected material that has broken down is not yet a complete compost: it becomes compost only through the addition of manure or a suitable substitute that supplies nitrogen, plus additions of phosphate, potash rock and bone meal.

Q. How can I judge the success of my compost pile?

A. The following are checkpoints with which to gauge the success of a compost heap:

a) The material should be medium loose, not too

60

tight, not packed and not lumpy. The more crumbly the structure, the better it'is.

b) A black-brown color is best; pure black, if soggy and smelly, denotes an unfavorable fermentation with too much moisture and lack of air; a greyish, yellowish color indicates an excess of dead earth.

c) The odor should be earthlike. Any bad odor is a sign that the fermentation has not reached its final goal and that bacteriological breakdown processes are still going on. A musty, cellar-like odor indicates the presence of molds, sometimes also a hot fermentation which has led to losses of nitrogen.

d) A neutral or slightly acid reaction is best. Slight alkalinity can be tolerated. One has to keep in mind that too acid a condition is the result of lack of air and too much moisture. Nitrogen-fixing bacteria and earthworms prefer the neutral to slightly acid reaction. The pH range for a good compost is therefore 6.0 to 7.4, 7.0 being neutral.

e) On the average, an organic matter content of from 25 to 50 per cent should be present in the final product. This means that from one- to two-thirds of the original material ought to be organic matter. The balance should be made up of earth, old rotted compost and lime.

f) Maintain a moisture content like that of a wrung-out sponge: no water should drip from a squeezed sample. However, do not let the compost get dry.

Q. How can I build an inexpensive compost bin?

A. The best compost bin is the one that turns your homestead's organic wastes into soil fertility quickly, cheaply and easily.

One gardener we know built one using scrap two-inch-square lumber, which he covered with half-inch chicken-wire mesh for a total cost of $4.00. Made of two L-shaped sections held together with screen-door hooks, the cage provided him with 18 to 24 cubic feet of finished compost in 14 days!

Cement-block construction is another way of building a rugged, durable compost bin. The cost of used blocks is low, and a double bin with a center partition can be made quickly and with little effort.

Q. What is the New Zealand bin?

A. The classic among compost bins is the wooden New Zealand box which is designed to admit as much air as possible from all sides. There are many variations of this box. Wooden slats, spaced far enough apart to allow moving air, are most popular as a material

A good example of a cement-block compost bin.

for making the New Zealand box.

Q. Can compost be made in a garbage can?

A. A garbage can makes an ideal composter. Just punch holes in the bottom and in the sides, as well as in the lid. Place about four inches of fertile soil in the bottom of the can. Then layer your materials in the same way as for a large heap in the garden. If you like, plants can be grown and trained over the can so that it appears to be a mound of green leaves.

Q. How can you be sure of getting air into the center of a compost pile?

A. There are several methods. One is turning the compost. An open lattice support which can be built into

This New Zealand-style compost bin is a popular way to make compost.

the bin and allows the air to pass under the compost is, perhaps, the most practical.

Application

IV

Q. When should compost be applied?

A. The principal factor in determining when to apply compost is its condition. If it is half finished or noticeably fibrous, it could well be applied in October or November. By spring it will have completed its decomposition in the soil itself and be ready to supply growth nutrients to the earliest plantings made.

Otherwise, for general soil enrichment, the ideal time of application is a month or so before planting. The closer to planting time the compost is incorporated, the more it should be ground up or worked over thoroughly with a hoe to shred it. A number of garden cultivating tools and machines offer an excellent time-and-labor saving hand in accomplishing this and will help spread the compost evenly and mix it with the soil.

If the compost is ready in the fall and you do not intend to use it until spring, keep it covered anp stored in a protective place. Water the finished compost from time to time if it is kept for a long period during the summer.

Q. How much compost should be applied to a garden annually?

A. Try to figure on applying a depth of one-half inch to three inches of compost over your garden each year. You can apply compost either once or twice a year. The amount depends, of course, on the native fertility of your soil and on what and how much has been grown in it.

Q. What about using compost in sowing seeds?

A. For sowing seeds indoors or in a cold frame, put your compost through a one-half-inch sieve, then shred it with a hoe or even roll it with a rolling pin to make it very fine. Then mix it with equal amounts of sand and soil. The ideal seeding mixture is very mellow and tends to fall apart upon being squeezed in your hand.

Q. What's the best way to apply compost in a flower garden?

A. Finely screened compost is excellent to put around all growing flowers. Apply it alone as an inch-thick mulch to control weeds and conserve moisture or top dress it mixed with soil. In the spring you can loosen the top few inches of soil in your annual and perennial beds and work into them an equal quantity of compost.

Use compost generously when sowing flower seeds. Compost watering is an excellent way to give your flowers supplementary feeding during their growing season. Fill a can half full of compost, add water and sprinkle liberally around the plants. The can may be refilled with water several times before the compost loses its potency.

Q. How should compost be applied in a vegetable garden?

A. To enjoy superbly delicious, healthful vegetables, apply compost, compost, compost. Dig it in in the fall, bury it in trenches, put it in the furrows when planting and in the holes when transplanting. After the plants start shooting up, mix it with equal amounts of soil and use it as a top dressing, or mulch them heavily with partially rotted compost or with such raw compost materials as hay, straw, sawdust, grass clippings and shredded leaves.

Q. How should compost be used on a lawn?

A. Want a lawn that stays green all summer, has no crab grass and rarely needs watering? Then use compost liberally when making and maintaining it. You want a thick sod with roots that go down six inches, not a thin, weed-infested mat lying on a layer of infertile subsoil. In building a new lawn, work in copious amounts of compost to a depth of at least six inches. If your soil is either sandy or clayey (rather than good loam), you'll need at least a two-inch depth of compost, mixed in thoroughly, to build it up. The best time to make a new lawn is in the fall. But if you want to get started in the spring, dig in your compost and plant Italian ryegrass, vetch or soybeans, which will look quite neat all summer. Then dig this green manure in at the end of the summer and make your permanent lawn when cool weather comes.

To renovate an old patchy lawn, dig up the bare spot about two inches deep, work in plenty of finished compost, tamp and rake well, and sow your seed after soaking the patches well.

Feed your lawn regularly every spring. An excellent practice is to use a spike tooth aerator, then spread a mixture of fine finished compost and bone meal. Rake this into the holes made by the aerator. You can use a fairly thick covering of compost—just not so thick that it covers the grass. This will feed your lawn efficiently and keep it sending down a dense mass of roots that are unaffected by droughts.

Q. What are the best methods for using compost on trees and shrubs?

A. When planting trees and shrubs always make the planting hole at least twice the size of the root ball in all directions. The best planting mixture is made up of equal parts of compost, topsoil and peat moss or leaf mold. Fill this in carefully around the plant, tamping it down as you put in each spadeful. Soak the ground well, then spread an inch or two of compost on top. A mulch of leaves and hay will help keep the soil moist and control weeds. Established

Ring Method.

shrubs should be fed yearly by having a half bushel of · compost worked into the soil surface, then mulched. When hilling up the soil around your rose bushes for winter protection, mix in plenty of compost—the bushes will get a better start next spring.

The "ring" method is best for feeding trees: start about two feet from the trunk and cultivate the soil shallowly to a foot beyond the drip line of the branches. Rake an inch or two of compost into the top two inches.

The ring method is ideal for fruit trees, too. You can work in as much as three or four inches of compost, then apply a heavy mulch which will continue to feed the trees. Some gardeners merely pile organic materials as the covering decomposes. You can even add earthworms to speed the transformation to humus. Berry plants may be treated the same way, with lower mulches, of course, for low-growing varieties. Another good trick for pepping up old fruit trees is to auger holes a foot apart all around the tree, and pack these with compost.

Do's and Don'ts

V

Q. Do we need "compost activators?" What do you think of them?

A. Compost activators are usually described as high-powered, bacteria-saturated substances which stimulate biological decomposition in a compost pile. We don't recommend them, nor do we think they are needed.

In our experience these additives don't help a well-made compost heap break down any faster. And for a poorly-made pile of organic materials—or for one made up mainly of straw, sawdust and weeds—it would do more good to add high-nitrogen materials like cottonseed meal, blood meal or manure along with any natural mineral fertilizers. Large-scale waste composting research has recently supported our opinion.

Q. Are coal ashes good material for a compost heap?

A. Although coal ashes will lighten heavy soil, there is

a real danger of adding toxic quantities of sulphur and iron from this material. Some ashes do not contain these chemicals in toxic quantities, but the coals from various sources are so different that no general recommendations can be made. The safest procedure is to regard all coal ashes as potentially injurious to the soil.

Q. Are diseased wastes safe to use for compost?

A. Diseased organic matter is suitable for composting provided the heap is properly made and tended. The high internal temperature created in the decomposing waster (about 160°F.) will destroy most common plant disease organisms. The important thing is to be sure the compost heap is built up carefully and that it includes a good balance of organic materials. It is advisable to keep a sizable ratio between uninfected wastes and diseased matter going into the compost. Turn the heap thoroughly after it heats so that plants near the surface layer are placed well inside where peak heat is reached.

Q. Redwood shavings and sawdust are plentiful on the West Coast. Are either of these detrimental to domesticated earthworms?

A. Well-seasoned redwood sawdust or shavings will not cause difficulty. Raw, green redwood sawdust or chips can be harmful. Like eucalyptus, they resist decomposition much longer than wood wastes from other trees and should not be incorporated into the soil. The presence of oils in the redwood debris might also be detrimental to other soil organisms. Using well-seasoned redwood in the compost heap or for sheet composting is best. Strawberries, rhododendrons, azaleas and other acid-loving plants do well in it.

Miscellaneous

VI

Q. Why insist on composting over the present methods of garbage disposal?

A. The three current methods of garbage disposal are incineration, landfill and dumpings: each has its disadvantages. The sanitary landfill often requires long-distance hauls to suitable sites; suitable sites are scarce in heavily populated areas; winter conditions can make operations almost impossible; future use of fill sites is limited; lack of available cover material may prohibit use of the method. Dumping has the obvious disadvantages of health and fire hazards. The incineration method presents air pollution problems, difficulties in salvaging useful material, waste of potential organic matter, high capital cost, high operation and maintenance costs, and high skills required for successful operation.

Q. What are the advantages of municipal composting in place of the current garbage disposal methods?

A. Composting has the following clear advantages over the other three:

a) Converts solid waste to a usable end product.

b) Compared with the sanitary landfill method, a centrally located compost plant reduces hauling costs which approximate one to three dollars per ton. Furthermore, getting rid of the noncompostable residue from the composting operation is less costly than the expenses involved with landfill.

c) Salvaging metals, rags, etc., is more economical in a composting operation because of the plant's design.

d) A well-designed compost plant can handle dewatered sewage solids, especially if the solids are mixed with ground refuse. (Cost of treating sew-

Municipal Composting Plant.

age solids by composting is about half the cost of conventional disposal methods in a modern sewage plant.)

e) Composting principles used for garbage and trash disposal also apply to any industrial wastes, so that a municipal plant could be used for the combined disposal of these wastes.

Of course, another basic advantage to composting is the fact that use of compost by gardeners and farmers would be increased tremendously, thereby aiding in the conservation of our nation's soil.

CHAPTER 5

EARTHWORMS

A bounty of worms will make a welcome difference in the soil.

Any serious gardener quickly learns that the earthworm is a valuable, useful friend. A large population of earthworms can enrich the mineral value of the soil, aerate it and even provide avenues for the rain to penetrate. Follow the rules for encouraging a large earthworm population: avoid DDT, lead arsenate and frequent applications of the chemical fertilizers that pose serious threats to the worms. Many successful gardeners find earthworms so desirable that they buy them and add them to their soil. It is an inexpensive, no-work way to do your soil a great favor.

Q. What role does the earthworm play in soil fertility?
A. The earthworm is a valuable adjunct in the soil's expression of fertility. He digests the soil—eats it

EARTHWORMS

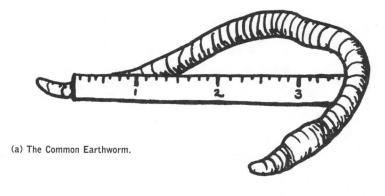

(a) The Common Earthworm.

(b) This profile shows how earthworms burrow to aerate soil.

and conditions it. To an important extent our top-soils have practically been made by earthworms: that is why Aristotle called them the intestines of the soil. Their castings are far richer minerally than the soil which they ingest. It is said that an average earthworm will produce its weight in castings every 24 hours. They burrow into the ground as far as six feet down, aerating the soil and making holes for rain to penetrate. Earthworms also break up hard-pans. Each year their dead bodies furnish a considerable amount of valuable nitrogenous fertilizer which may amount to more than a thousand pounds per acre in a highly "organic" soil.

73

Q. What can be done to encourage a large earthworm population?

A. First of all, the farmer who wishes his crops and soil to benefit from the work of nature's champion composter must avoid the use of all chemical fertilizers, sprays and dusts. These are fatal to worms. Nitrogenous fertilizers, for example, because they tend to create acid conditions, wipe out earthworms rapidly. Dead worms are found on top of the soil in huge numbers whenever a chemical fertilizer is applied, and one garden magazine went so far as to suggest that frequent applications of a chemical fertilizer are the quickest and easiest way to get rid of worms.

In Australia the use of superphosphate on pastures almost totally wiped out even so hardy a specimen as the giant 9-foot-long Gippsland earthworm. DDT, toxaphene, lime, sulfur, mercuric chloride, lead arsenate and a host of others have been shown to be deadly to earthworms.

As far as increasing the earthworm population is concerned, the best and only proven method has been the application of organic manure, a source of food for the earthworms which favors their development. Two British researchers, Sears and Evans, have found that the addition of organic matter to the soil not only increased the number of worms per acre, but also brought about an increase in the size of the earthworms.

Q. Which type of earthworm is best for use in a compost pit—domesticated, hybrid, red wrigglers or brown-nosed angle?

A. The first three varieties mentioned are all manure worms. The brown-nosed angle worm is the regular type of night crawler. In order for manure worms to thrive, they must live in manure. Sometimes they do well in rich compost that has as much nitrogen as manure. In general, manure worms are best for use in compost heaps. They can be raised either indoors or outdoors, an advantage in the North. Some Florida worm breeders, however, have switched to angle worms or night crawlers. They are larger and more accustomed to working in actual soil. And they are preferred by fishermen.

Q. How should a compost pit for earthworms be constructed?

A. Plan to plant 500 breeders or 1000 mixed sizes per cubic foot of compost. They will do their job quickly and well. Be sure to have a solid, or a one-half-inch wire netting bottom on your pit to keep the moles from ravaging it. Planking, shiplap, or a thin layer (one inch) of concrete will do. The sides can be similar wood, four-inch concrete walls, or concrete blocks. An over-all height of two feet provides room for ten to twelve inches of compost on the bottom and protective layers of dry leaves, wilted grass, weeds, green garbage, or hay to shield the worms from the elements.

Q. When adding earthworms to a compost pile, does it matter where we put them?

A. The worm works best in the upper six to ten inches. It's best to keep the compost pile low enough for him to do his best if space permits.

Q. What is the best way to remove finished compost from a pit without removing the earthworms with the compost?

A. Strike the top of the compost in the pit with a stick or shovel and the earthworms will move towards the bottom of the pit. The compost in the upper part of the pit may then be removed without taking earthworms that are old enough to move downward.

Q. Is salt used in a garden to kill slugs injurious to earthworms?

A. There is no place in the garden for salt. It is used sometimes as a weed killer, but the amount of salt necessary in order to be effective, would be injurious to all plants and soil bacteria. To discourage slugs we suggest that you spread fine gravel over the soil as slugs avoid rough surfaces. If your soil is fine humus it should normally have an earthworm population. The applications of salt may drive them away. If salt has already been applied, try planting a fresh supply of worms after a time, in the hope that most of the salt has leached away.

Q. What happens to earthworms during the winter?

A. During the time that the ground is frozen, earthworms migrate below the frost line. The activity of the earthworms is then considerably decreased. But even if earthworms are frozen in the soil, they do not die.

Q. How late in the fall can I add earthworms to the soil?

A. You can add earthworms to the soil until the ground is frozen, but be sure to protect the ground with a mulch or cover crop.

Q. Is a power gardening cultivator harmful to earthworms?

A. It may kill a few but their dead bodies will enrich the soil. These cultivators do such a wonderful job of thoroughly distributing compost in the soil that they are highly recommended.

Q. If earthworms are cut with a spade do they die? It is almost impossible to avoid cutting a good many.

A. Earthworms that are cut with a spade sometimes recover, that is one part of them may recover. The other part will die. However, there is not a total loss because their decaying bodies add nutrients to the soil.

Q. Can pure earthworm castings be used as compost?

A. Earthworm castings are the most fertile kind of compost. This compost can be used pure for seedlings, but it is a practice to use $\frac{1}{4}$ castings and $\frac{3}{4}$ good soil.

CHAPTER 6

WEEDS, SEEDS AND SEEDLINGS

Soil temperature, moisture and depth of planting are just a few things that must be considered in getting seeds to produce their best yield. The organic gardener who wants to be positive that the seeds he plants are really organic and have not been treated with preservatives or fungicides (a common commercial practice) dries them himself. If this is not feasible, there are suppliers who will provide untreated commercial seed on request.

Start seedlings indoors, before planting time, then transfer them to the garden, so you can enjoy the fruits of your labors earlier and longer. But remember, healthy seedlings depend on a healthy soil. The answers in this section will help you avoid such threats to seedlings as fungus infestation.

Weeds are the scourge of the gardener. They can prevent seeds from sprouting, choke out struggling seedlings and even kill mature plants. Too many gardeners, tired of digging up weeds and pulling them out, turn to chemical weed killers only to find that what kills weeds tends to kill flowers, shrubs and vegetables as well (not to mention what chemicals do to valuable soil bacteria). The following pages show that it is indeed possible, even easy, to grow healthy plants and to eliminate persistent pesty weeds by natural methods.

Eliminating Weeds

I

Q. How can unwanted plants be eliminated in a garden without using weed killers?

A. Start by pulling them out or digging them up. Then cover the entire surface with a smother mulch, which may include waste paper or cardboard, and keep up continuous cultivation to cut down top growth and eventually kill underground roots. Where livestock and poultry may be kept, grazing goats, ducks or geese will reduce edible weeds quickly.

Dandelion

Q. Is there a safe way to get rid of large patches of well-rooted poison ivy?

A. Large infestations of poison ivy can be controlled by mowing close to the ground in midsummer, followed by plowing and harrowing, or by grazing sheep or goats. Small patches can be grubbed out. Under trees or along a fence where plowing may be difficult, mow regularly by machine. Where plowing is not feasible, try smothering the ivy with tar paper or cardboard. A deep mulch of straw or other material will also be effective—and more attractive.

Vines growing up fences or in trees can be handled by cutting each one near the ground and pulling the wilted plants from the fence or trees several

Giant Ragweed

days later. Wear gloves and protective clothing. When finished, wash well with any yellow soap, such as Fels Naptha.

Q. What is the best way to eliminate quack grass?

A. This is a difficult weed to control because it produces so many underground stems, each joint of which is capable of growing into a new plant. Cutting up a plant with a cultivator or hoe is in effect propagating the plant asexually. All rootstocks which are

Rough Pigweed

hoed or pulled out should be composted in a compost heap. By persistent and continuous hoeing and pulling, quack grass can be kept under control.

Quack grass cannot stand crowding by other plants or shade. By feeding the lawn grasses early in spring, the grasses will grow with such vigor that the quack grass and other weeds will be crowded out.

Q. Do you ever recommend the use of chemicals to kill weeds?

A. No. Chemicals which kill weeds are apt also to kill the beneficial organisms in the soil.

Q. How can crabgrass be eliminated from a lawn?

A. Feed your lawn plenty of organic nutrients. Don't cut it lower than three inches with the mower. Give it a good drenching from time to time. You'll get rid of crabgrass.

Seeds and Seedlings

II

Q. What is damping-off?

A. A disease caused by fungi that kills many young plants and is characterized by stem collapse or seedlings falling over. It may occur before seeds germinate or after seedlings emerge. Disease infestation in roots or lower stems of seedlings results in a stunted, unproductive plant. In some cases the disease may affect the tops of seedlings and kill leaves and growing points with no damage to roots.

Q. How can damping-off be prevented?

A. Many of the fungi that cause damping-off are present in the soil. Potting or seed-flat soil to be used in plant production should be treated to kill any that might be present. This may be done by steam-heating the soil to 180 degrees F. for one-half hour or more. Since this is often impractical for the home gardener, dry heating in an oven will also kill fungi in the small amounts of soil used. Here's how: Soil should be four to five inches deep in the container(s)

81

with a small potato (1½ inches in diameter) buried in it. Place the soil in an oven at 200 degrees F. and heat until the potato is cooked. The soil is then sterilized and ready for use.

Q. What is vermiculite and why is it suggested for starting seedlings?

A. Vermiculite is a mineral belonging to the mica family (mineral silicates) which has rapidly gained acceptance as a medium for starting seedlings and root cuttings. When vermiculite ore is heated to about 2,000 degrees F. in processing, the moisture in it turns to steam, popping the granules to many times their original size. Countless tiny air cells thus produced provide additional air- and water-holding capacity, which is an aid to germination and the development of dense root systems. Vermiculite can hold water in amounts several times its own weight. Even when thoroughly wet, ample air circulates around plant roots, helping to avoid damping-off.

For mulching or as a soil conditioner to lighten and aerate heavy clay soils and to help sandy soils retain moisture, vermiculite is valuable; but it does not perform these jobs as well as does organic matter, which also feeds plants and brings other benefits. Vermiculite can also be used in storing bulbs and winter vegetables, and as a base for floral arrangements.

Q. What is meant by stratification of seeds? Why is this done with some types of seed and not others?

A. Stratification is a method used for preparing hard-coated seeds for planting because germination is generally very slow. Seeds from most fruit or nut trees, for example, require a certain amount of winter chilling before they will germinate. In order to save time in getting such seeds to germinate, they are layered in sand, sawdust or peat moss and then exposed to freezing and thawing for a period of time sufficient to duplicate natural winter conditions. When starting tree seeds in late winter or early spring, the stratification process can be speeded up by placing the seeds, mixed with moist sand, in a deep freeze for a few days. This freezing and thawing process is repeated for a week or two before the seeds are planted. When more time is available, the stratification box is placed in a refrigerator or country root cellar for 30 to 120 days, depending on the type of seed being germinated. If a refrigerator is used, it is advisable to place the box on a bottom shelf with a thick piece of cardboard under it to absorb moisture.

Q. Directions on the seed packets of most garden vegetables simply include advice to "cultivate frequently" during the growing season. Why is such "cultivating" necessary and what does it do for the plants being grown?

A. Two types of cultivation are important to successful gardening. Before planting, cultivation refers to preparing the soil—lightly digging the top several inches to loosen the earth, break up clods, and help aerate the growing site. For small areas, most gardeners use a spading fork or a garden spade in sandy soils. When turned and loosened, the planting location should be raked fairly smooth, to a depth of two to three inches. Once seed is sown or plants set out, the main purpose of cultivating is to prevent competition between crops and weeds for both water and fertilizer. This cultivation should be quite shallow—kept to about an inch deep—because damage to crop roots can easily result from deeper working.

On soils which tend to crust, cultivation is often needed to permit water to penetrate more rapidly. Cultivating during the growing season can be eliminated by using various mulches—hay, straw, leaf mold or ground corn cobs. Besides being an effective method to keep down weed growth and to help retain moisture, mulching also enriches soil as it decomposes.

Q. Does the protein in sprouted seeds become a complete protein?

A. Soybeans contain protein of a quality similar to that of meat. Other seeds do not. That is, they are lacking or deficient in one or another of the amino acids, or forms of protein. Sprouting the seed does not improve the quality of its protein, although it does increase its vitamin content.

Q. How can the highest percentage of germination be obtained, from all seeds, with a minimum of time and effort?

A. 1. Do not sow seeds outdoors in the garden until the

Start seeds indoors in areas where growing season is short.

84

soil has warmed enough for the type being sown. Basic observation reveals that *all* seeds sprout more promptly in summer than in spring. This is true even of the kinds (peas, lettuce, etc.) that prefer cool weather for growth of the plants.

2. Try to have the top inch of soil especially rich in humus. This insures an adequate moisture supply without rain or using the hose. It also insures a loose soil structure for easy emergence of seedlings.

3. When sowing seeds, do not cover them any deeper than necessary. Seeds need light and oxygen as well as moisture for germination.

4. When seeds are covered, firm the soil to give them intimate contact with moist (*not wet*) soil. This is more important than any amount of future watering.

5. If the soil is moist when seeds are sown, no future watering is necessary—except for rare kinds that take 20 days or more to sprout. If soil is dry, give the seedbed a soaking with a *gentle* spray from the hose. Better yet, make the drills for seeds, soak these with water, press seeds into the wet soil, then cover with barely moist soil. Either way, the single watering is enough to insure germination. More may do more harm than good.

6. While seeds do not require an abundance of moisture for sprouting, seedlings do require it. Once sprouting has begun, give the seedbed (or garden) a thorough soaking. Even then, however, do not keep soil too wet. The ideal is to have the root zone moist but the surface relatively dry. This encourages the deep rooting that results in more vigorous plants with much better resistance to drought.

Q. Should seeds that have reached maturity be used for the next planting?

A. Seed that is to be saved must be ripe, even a little

overripe. If rained on, seed should dry out again before being gathered.

Q. After seeds have been gathered, what is the best method for drying and storing them?

A. Hang seed stalks in a dry, airy place (shed, cellar, attic) until they're brittle-dry. Hang up vines of peas and beans, or strip off pods and lay seeds on racks (not on the floor as dampness will rot the seeds). Hang up ears of corn by the braided husks.

Q. How are seeds from soft-fleshed fruits and vegetables saved?

A. Tomatoes, cucumbers, melons and other soft-fleshed vegetables and fruits are treated differently. Scrape out the center pulp from a dead-ripe tomato. Place it in a saucer with a little water and let it ferment a day or two. Then sieve it all and wash out the pulp. Dry the remaining seeds on absorbent paper.

Handle all seeds carefully. Nicked or injured seeds won't keep well. Store large seeds in glass jars with loosely fitting (not airtight) covers; keep small seeds in paper envelopes. Label seeds and mark the date. Most seeds—if they were good to start with and were gathered ripe and kept dry—are guaranteed viable for at least two years. Some seeds like melon, squash and cucumber last for 8 or 10 years.

Q. How do seeds in general compare with other foods?

A. They can provide a valuable part of the well-balanced diet. Seeds abound in essential trace minerals and other nutrients that are considerably less plentiful in many other plant foods.

CHAPTER 7

TREES AND SHRUBS

When you invest in a tree or expensive shrub, you want to keep it flourishing through its first insecure months in a new soil. Protecting young trees from intense sunlight that could burn them out and providing them with proper drainage are services basic for their survival. Keep trees well nourished and you will discourage many destructive arboreal insects. These essentials can be accomplished without resorting to chemicals. There are simple ways to insure the survival of evergreen seedlings and shade trees. Learn the right way to plant and protect new trees and to prune and fertilize older ones and you have mastered some major mysteries of successful landscape planting.

Q. What is the best method for planting trees?

A. Plant trees either in early spring before new growth begins, or in the fall when they become dormant. Here are planting tips:

a) Soak the roots in water for 12-24 hours before planting.

b) Dig the planting hole wide enough for roots to spread out. A few long roots can be trimmed back to fit the hole, but avoid excessive pruning. Any injured roots should be pruned.

c) Make the hole deep enough for trees to set one to two inches deeper than they grew in the nursery row. (Dwarf varieties of fruit trees shouldn't be set this deep; the graft union must be kept above ground.)

d) When filling the hole, put fertile top soil around roots and tamp lightly. Once roots are covered,

87

Plant deep enough and wide enough to accomodate tree's roots.

pack soil firmly with the heel of your shoe to avoid air pockets. Soil from the bottom of the hole may be used to finish filling. Tamp lightly when completed.

e) If the trees are branched, cut off branches that make sharp angles with the trunk. Leave one-to-three well-spaced lateral branches to grow into main scaffold branches.

88

Set tree one or two inches deeper than it grew before transplanting.

    f) Apply mulch immediately after planting to conserve soil moisture and aid new root growth. If you use straw, apply one-half to one bale per tree.

Q. How should fruit trees be fertilized?

A. Use fish fertilizer annually to build up minor elements in the soil. Start with one teaspoonful to five gallons of water concentration at planting and increase one teaspoonful each year, up to a quart

Fill in hole with fertile soil.

for full-size trees (half that much for dwarfs).

Apply in three stages: one-third in early spring when trees are dormant and before buds emerge; one-third after blossoms fall; and one-third in early summer to boost fruiting.

When it comes to solid fertilizers, apply 100 to 200 pounds of compost or leaf mold per tree in a two-to-three-foot circular band around the drip

Cut off branches that make sharp angle with trunk.

line. Spring is the best time for an application.

Experts may vary in their advice on how much to apply, but, generally, they agree that organic fertilizers are best for your trees.

Q. How may chemical sprays be eliminated in growing trees?

A. If the trees have been fed liberally with compost for a number of years, they should be relatively re-

Water newly planted tree.

sistant to diseases and insects. It will require some years to build up your trees to this point. In the meantime take protective measures that will not further injure the soil.

Spray trees early in the spring, before any of the buds open, with a three per cent oil spray to kill scale insects and insect eggs. Oil sprays may be purchased in most agricultural supply stores. This

92

dormant spray kills the insects by forming a film over their bodies which excludes oxygen. Sprays containing sulfur, copper, arsenic and the highly toxic sprays and dusts should be avoided. Swabbing tree trunks with tanglefoot (a commercially available sticky substance) will catch moths or worms so that they cannot get into the bark. Planting garlic cloves close to the tree trunk has been found effective against the borer as has the use of trichogramma wasps.

Q. Is it worthwhile to reclaim old, unproductive fruit trees?

A. If the trunk or branches are badly rotted, or if a quarter of the top is dead through disease or winter injury, it is rarely worthwhile to attempt salvage. However, here are some general steps to follow when bringing new life to old, neglected trees:

Cut out old wood and prune heavily to strong new growth; remove all suckers that are not necessary to

Well-fed trees produce a good crop, weather conditions permitting.

93

replace the top; prune out interlacing branches to open the trees to light and the circulation of air; break up the soil around the tree, working in a great deal of compost, manure and other organic materials; apply organic nitrogen to the soil in the form of dried blood, cottonseed meal, or nitrogen-rich sludge—about 25 to 35 pounds per tree; mulch heavily. Do this regularly for several seasons.

Q. Can canker worms be discouraged without resorting to undesirable poisons and chemicals?

A. It is the female canker worm who causes most of the trouble by laying eggs. Fortunately, the female of this species does not have wings; she must crawl up the trunks in order to reach a satisfactory place to lay eggs. Since there are two type of canker worms— the fall and spring canker worm, a protective sticky band placed around the tree trunk at the start of each of these seasons will eliminate many of the insects. The sticky bands must be fresh, with no gaps or bridges for the insects to cross. One banding material is Tree Tanglefoot. Other products available include aerosol cans of an adhesive compound in liquid form for easy spray-on application. Trik-O is also effective in controlling this pest.

Q. What can be done to combat fire-blight on pear trees?

A. There are ways of controlling this very serious enemy of pears that are successful in many areas. First, plant varieties that are particularly resistant. We have had success on the Organic Gardening Experimental Farm with Kieffer, Bosc and Seckel. Then be sure not to fertilize young trees heavily. The tender, new growth of fertilized trees is most appealing to the blight. By giving them hard conditions for the first few years, at least, you may get hardier trees. Finally, if the blight does strike, trim off the diseased parts as fast as they appear.

Burn cuttings at once and sterilize tools thoroughly after each pruning.

Q. What measure should be taken against peach leaf curl?

A. Leaf curl of peach is caused by a fungus which spends the winter on the buds of the trees. You should practice sanitation and be sure to spray your trees while dormant with Scalicide. This is a 3 per cent oil spray which is not injurious to the tree or soil organisms. As soon as new leaves show curling, they should be removed to the compost heap. By watching your trees carefully you should be able to reduce the damage of leaf curl considerably. Fruit trees seem to become more resistant to diseases and pests when they are well fertilized with compost.

Q. Is whitewashing tree trunks worthwhile?

A. Yes, especially in certain regions of the country. Among the advantages claimed:

a) Whitewashing reduces insects and disease organisms that burrow in the bark tissues; insect eggs are presumably killed by the whitewashing substance.

b) Whitewashing prevents some scald and heat injury of the bark tissues of very young trees. This is especially true in the sunny regions of the country.

Generally, the whitewash is applied over the trunk and main supportings of the tree.

Q. Why are the trunks of newly transplanted trees sometimes wrapped?

A. Young trees set out in lawns or orchards often are given some sort of wrapping to protect them from drying out and from sunscald or windburn. Wrapping also helps prevent damage from borers and some other pests. Burlap and double-thick paper are materials popularly used. Ordinary aluminum foil does the job too, adding protection against trunk girdling by rabbits.

95

Wrapping trunk of newly transplanted tree.

Q. Is there a simple way to determine if a yard has proper drainage for planting trees?

A. A good way to test for poor drainage is to dig a hole about 18 inches deep where you plan to plant a tree. Fill the hole with water and allow it to stand for 36 hours. If water remains in the hole at the end of that time, drainage is unsatisfactory. Build up the soil by turning in compost or other rich organic matter, or, if time permits, try growing a winter cover crop to improve its structure.

Q. How should evergreens be fertilized? Is ordinary compost enough?

A. If they are dwarf junipers and pines they will require none unless the soil is very poor. Arborvitae, yews and chamaecyparis like some fertilizer. The best way to apply nourishment to established plants is to use a top dressing mixture of leaf mold or peat moss and old manure before the snow falls. They can be fed later in the winter by applying a rich mulch of two to three inches of manure or leaf mold.

Q. How should evergreens be protected against winter-kill?

A. The evergreens take their moisture from the soil. From the beginning of October until mid-November, or until the snappy weather arrives, make certain that the ground receives thorough waterings at least every other day. Then, after the ground has frozen at least six to eight inches deep, apply the winter mulch.

Remember, the object is not to prevent root freezing but to prevent freezing and thawing on top. Any hardy plant can stand freezing, but it is likely to winter-kill if the surface of the ground thaws in the daytime, turns into sloppy mud, and then freezes hard again at night.

Q. How should seedlings of evergreens intended for ornamental use be planted?

A. Some soil preparation will be helpful. Dig a hole about 12 inches deep and 12 inches wide; fill it with a mixture of good soil plus about a handful of cottonseed meal and some leaf mold from an area where evergreens are growing wild. The cottonseed meal is a complete, organic plant food; the leaf mold usually contains certain fungi, known as mycorrhiza, which aid in proper growth of pines and other evergreens. If you have had difficulty in establishing pines, it may be helpful to add some of this leaf mold to each spot where a tree is to be planted.

Q. What suggestions do you have for fertilizing evergreens and shade trees?

A. One method which has been highly successful is a rock dust mixture. It consists of one part ground limestone, two parts phosphate rock, three parts granite dust, three parts cottonseed meal and four bushels of compost. This is applied at the rate of one pound for each one foot of drip-line diameter. If the tree measures 15 feet tip to tip of opposite branches then 15 pounds should be applied from tree trunk to drip line. A three-inch mulch completes this annual treatment. Fertilizer mixtures should be worked into the soil to encourage deeper root growth. This can be done by making holes one foot apart with a crowbar, starting two feet from the trunk and extending to the drip line. The fertilizer should be placed in each of the holes.

Q. When can shrubs or trees that have been grown in containers be planted out of doors?

A. Almost any time during the year that your soil can be readied. Remove the plant from the container carefully, to avoid breaking the ball of soil. (Watering the soil first will help in this, or it may be necessary to cut the container.)

Dig a hole large enough for the ball of soil and backfill—about one foot larger than the ball. In a heavy clay soil, mix one-third by volume of coarse peat moss, partially rotted manure, or other organic materials. Firm soil and water thoroughly. Dishing the soil around the plant will aid water penetration. During a dry spell, new plants should be watered well every week to ten days.

Q. What can be done for shrubs that suffer from "winter drying?"

A. Winter drying is not a disease but a case of dehydration because the plant is unable to obtain sufficient moisture due to frozen or dry soil. Damage usually shows up after periods of rapid changes in temperature, especially if accompanied by drying winds. In such evergreen shrubs as the juniper and arborvitae, it results in browning and death of foliage, primarily on one-year old growth at the tips of branches. The best preventive is a good mulch placed around the trees in the fall and maintained throughout the winter. After damage has occurred, the only treatment is to prune off dead tips and feed the shrub to give it vigor for new growth. Some browning on interior limbs is natural on certain species of evergreens in the spring.

CHAPTER 8

MULCHING

Mulching is a method organic gardeners use to discourage weeds, preserve ground moisture, protect roots against freezing in winter and keep low-growing vegetables clean and dry. Mulching means covering the bare soil with organic material that can be anything that's ever been alive, from corn stalks to hay or leaves. Choose the mulching materials that are most easily accessible. You will find that almost any location offers a wide selection.

Follow the directions for the thickness of the mulch application in different projects and pay particular attention to how the ground should be prepared before mulching. Caution should be observed in mulching, since it encourages a high humidity that is not always desirable. However, careful mulching has been a lifesaver for many gardeners; it can be for you too.

Pros and Cons

I

Q. What is mulching?

A. As with composting, mulching is a basic practice in the organic growing method; it is a practice which nature employs constantly, that of always covering a bare soil. In addition, mulching also protects plants during winter, reducing the danger of freezing and heaving. So a mulch is defined as a layer of material, preferably organic material, that is placed

Protect trees from harsh weather by thick mulch of straw and leaves.

on the soil surface to conserve moisture, hold down weeds, and ultimately improve soil structure and fertility.

Q. What are the advantages of mulching?

A. Experiments have shown the following advantages:

a) We know that a mulched plant is not subjected to the extremes of temperatures that an exposed plant is. Unmulched roots are damaged by the heaving of soil brought on by sudden thaws and sudden frosts. The mulch acts as an insulating blanket, keeping the soil warmer in winter and cooler in summer.

b) Certain materials used for a mulch contain rich minerals, and gradually, through the action of rain and time, these work into the soil to feed the roots of the plants. Some of the minerals soak into the ground during the first heavy rain. Therefore, mulch fertilizes while it is on the soil surface as well as after it decays.

101

c) For the busy gardener, mulching is a boon. Many backbreaking hours of weeding and hoeing are eliminated. Weeds do not have a chance to get a foothold and the few that might manage to come up through the mulch can be hoed out in a jiffy. And since the mulch keeps the soil loose, there is no need to cultivate.

d) The mulch prevents the hot, drying sun and wind from penetrating the soil; therefore, soil moisture does not evaporate as quickly. A few good soakings during the growing season will tide plants over a long dry spell. It also prevents erosion from wind and hard rains. Soil underneath a mulch is damp and cool to the touch. Mulched plants often endure a long dry season with practically no water.

e) At harvest time vegetables which sprawl on the ground such as cucumbers, squash, strawberries and unstaked tomatoes, frequently become mildewed, moldy or even develop rot. A mulch prevents this damage by keeping the vegetables clean and dry. During the harvest season, when most gardens begin to look unkempt, the mulched garden always looks neat and trim. In addition, mud is less of a problem when you're walking on mulched rows, and low-growing flowers are not splashed with mud.

Q. What are the disadvantages of mulching?

A. Here are some of the potential disadvantages of mulching:

a) Seedlings planted in very moist soil should not be mulched immediately. The addition of any organic matter which keeps the soil at a high humidity encourages damping-off of young plants. Allow seedlings to become established before mulching.

b) It is wise, too, to consider the danger of crown rot in perennials. This disease is also caused by a fungus. If there have been especially heavy rains, postpone mulching until the soil is no longer waterlogged. Do not allow mulches composed of peat moss, manure, compost or ground corn cobs to touch the base of the plants. Leave a circle several inches in diameter. The idea here is to permit the soil to remain dry and open to the air around the immediate area of the plant.

c) Do not mulch a wet, low-lying soil, or at most, use only a dry, light type of material such as salt hay or buckwheat hulls. Leaves are definitely to be avoided because they mat down and add to the sogginess.

Materials

II

Q. What materials can be used in mulching?

A. Practically any organic waste material can be used for mulching. However, since different materials have different textures and other properties, they differ in suitability. Here is a list of the most commonly used mulch materials.

Grass Clippings	Peat Moss
Shredded Leaves	Rotted Pine Wood
Cocoa Bean Hulls	Straw
Pine Needles	Sawdust
Weeds and Native Grasses	Corn Cobs
Salt Hay	Alfalfa Hay
Corn Stalks	Buckwheat Hulls
Unshredded Leaves	Stones
Cocoa Bean Shells	

Q. What determines the proper quantity of mulch that should be used?

A. During the growing season, the thickness of the mulch should be sufficient to prevent the growth of weeds. A thin layer of finely shredded plant materials is more effective than unshredded loose material. For example, a four- to six-inch layer of sawdust will hold down weeds as well as eight or more inches of hay, straw, or a similar loose, "open" material. So will one or two inches of buckwheat or cocoa bean hulls or a two- to four-inch depth of pine needles. Leaves and corn stalks should be shredded or mixed with a light material like straw to prevent packing into a soggy mass. In a mixture, unshredded leaves can be spread eight to twelve inches deep for the winter. To offset the nitrogen shortage in sawdust and other low-nitrogen materials, add some compost, soybean or cottonseed meal.

Q. Should fertilizer be used before mulching?

A. Before mulching fertilizer should be applied in the usual way. Compost, manures, rock powders and other organic materials will tend to decompose more quickly under a cooling, moisture-holding hay mulch; therefore surface fertilizing (or sheet composting) is very beneficial before mulching. Use lots of phosphate rock, granite stone meal (good for potash and mineral supply), and a magnesium limestone (only when needed to raise pH).

Q. Is there any value (or harm) in the use of coffee grounds as a mulch or mixed in with the soil?

A. Coffee grounds are of value as a soil conditioner. They contain as much as two per cent nitrogen, one-third of one per cent phosphoric acid and varying amounts of potash. Grounds sour easily because they preserve moisture well and seem to encourage acid-forming bacteria. Generally it is best to mix them with other waste materials for mulch.

Q. Are crushed peanut hulls suitable for use as a mulch in vegetable gardens and flower beds?

A. Peanut shells are excellent material for mulches. Besides the benefits of weed-control, improved moisture retention and soil cover gained through mulching, peanut hulls are surprisingly rich in nitrogen, containing 3.6 per cent which is much higher than straw. They also provide .70 per cent phosphorus and .45 per cent potash.

Q. How much sawdust mulch is necessary to keep down weeds in a vegetable garden?

A. In most instances, a one-inch layer of sawdust applied when vegetable plants are three to four inches tall will be enough to eliminate most of the need for cultivation, besides conserving soil moisture. Sawdust also can be used as a two-to-three-inch-wide, one-fourth inch deep band over rows after seeding. Mulch used this way will prevent soil crusting—one of the problems in gardening—but cultivation will be needed to control weeds between rows. A good supply of organic matter in your soil, or some nitrogen-rich materials like cottonseed or soybean meal, dried blood or bone meal used with the sawdust will prevent nitrogen shortage which sometimes develops while this mulch decomposes.

Q. How is sawdust used as mulch?

A. Sawdust and chips make an excellent mulch for apples, peaches, pears, and other tall fruit trees. Both are fine for blueberries, asparagus, strawberries and also for all kinds of shrubs, flowers and vegetables. In Canadian tests sawdust mulching increased the yield of raspberries by 50 per cent. In one very dry season, 1955, the yield of raspberries per acre was 3,850 pounds on sawdust mulch, 3,100 pounds on straw mulch, and only 1,742 pounds on clean cultivation. The mulch saved moisture, cut down weeds and raised the yields.

Apply a sawdust or chip mulch about three or four inches thick to the base of trees, or in rows. Regardless of whether you stake tomatoes or let them ramble, a mulch of chips or sawdust is useful. Staked tomatoes benefit since their foliage is off the ground and cannot shade the roots. Mulched to-

Tomatoes mulched with sawdust.

matoes get less blossom end rot—the leathery condition found on the bottom of tomato fruits in dry years. Unstaked tomatoes can be allowed to roam over a sawdust or wood-chip mulch, and in the fall the waste materials can be plowed under, after a liberal feeding of a nitrogenous fertilizer.

Q. Isn't there any disadvantage to using sawdust?

A. Nothing serious, aside from the temporary nitrogen tieup easily corrected. Sawdust may stick to damp strawberries and hamper early morning pickings or pickings following rain. This isn't a major problem, though, when you consider that sawdust or chips are free and help in so many other ways.

So if your soil is sandy or clayey, dries or cakes in hot weather, don't be afraid to use sawdust or chips for a mulch and conditioner. After all, sawdust and wood chips are composed of the same stuff leaves and woods earth are—and everyone knows the value of these materials.

106

Q. How about shredded bark as a mulch?

A. Fine, if it's free or you don't have to pay much for it. Bark has the same advantages that wood chips and sawdust have, usually looks attractive, needs extra nitrogen.

Q. What is peat? Are there different kinds?

A. Peat is formed by layers of successive generations of plants under water. In the absence of air, these plants decompose very slowly. Generally speaking, there are four major plant groups from which peat is derived: (a) the sphagnum group, composed of various herbaceous plants, has a strong acid reaction (pH 3.5 to 4.5); (b) the hypnum group, associated with mosses, sedges and other flowering plants, is usually neutral or slightly alkaline; (c) the reed-sedge group, formed from grasses, sedges and cattails; and (d) the shrub and tree group formed from blueberries, alder or willow. This yields a peat which lacks uniformity.

Miscellaneous

III

Q. How can the decay or rotting of a large straw stack be hastened so that it is suitable for mulching or sheet composting?

A. A good way is to make holes in the straw stack with a rod and pour in considerable quantities of manure water or water filtered through a compost heap. Such water will supply the bacteria and nitrogen that the straw needs to decay properly and quickly.

Q. What is the best time to start mulching for protection against winter weather?

A. The best time to start making a protective winter covering for your garden is long before the garden year ends. Accumulate all the organic matter that

MULCHING

you can find from residue crop materials, manures, cut weeds or outside products. Then lay down a thick mulch over the area to be planted to crops next year. All plant residues should be strewn over your garden for winter cover unless they are seriously diseased. Otherwise, gather up the infected plant residues, compost them, scatter the compost on the soil, and rake it in.

CHAPTER 9

VEGETABLES

Among the most satisfying, and the most tangible result of gardening, is the bountiful harvest of a vegetable garden. It is helpful to know that certain plant combinations are particularly compatible and bolster each other in the garden, while a few others are mutually detrimental.

Vegetable gardens have particular problems and the following pages offer some valuable tips for coping with them. Before you start planting your vegetables, take a look at the schedule we have provided that suggests the best times of the year to plant particular vegetables and the recommended distances between the rows.

Eating fresh, organically-grown vegetables from your own garden is the surest way to avoid ingesting harmful pesticides.

Planting

I

Q. Is it true that planting vegetables in certain combinations can help assure a good crop and prevent some common troubles?

A. There are many garden plants that have a natural affinity for one another and tend to stave off pests and diseases. The "protective planting" or companion planting idea is a basic one in the Biodynamic farming method.

Q. Can you give some useful planting combinations for vegetables?

A. *VEGETABLES*   *COMPANION*

Asparagus	Tomatoes, parsley, basil
Beans	Potatoes, carrots, cucumbers, cauliflower, cabbage, summer savory, most other vegetables and herbs
Beans, Pole	Corn, summer savory
Beans, Bush	Potatoes, cucumbers, corn, strawberries, celery, summer savory
Beets	Onions
Cabbage Family (Cabbage, cauliflower, kale, kohlrabi, broccoli, Brussels sprouts)	Aromatic plants, potatoes, celery, dill, camomile, sage, peppermint, rosemary, beets, onions
Carrots	Peas, leaf lettuce, chives, onions, leek, rosemary, sage, tomatoes
Celery	Leek, tomatoes, bush beans, cauliflower, cabbage
Chives	Carrots
Corn	Potatoes, peas, beans, cucumbers, pumpkin, squash
Cucumbers	Beans, corn, peas, radishes, sunflowers
Eggplant	Beans
Leek	Onions, celery, carrot
Lettuce	Carrots and radishes (lettuce, carrots and radishes make a strong team grown together), strawberries, onion
Onion (including garlic)	Beets, strawberries, tomato, lettuce, summer savory, camomile (sparsely)

Parsley	Tomato, asparagus
Peas	Carrots, turnips, radishes, cucumbers, corn, beans, most vegetables and herbs
Potato	Beans, corn, cabbage, horseradish (should be planted at corners of patch), marigold, eggplant (as a lure for Colorado potato beetle)
Pumpkin	Corn
Radish	Peas, nasturtium, lettuce, cucumbers
Soybeans	Grows with anything, helps everything
Spinach	Strawberries
Squash	Nasturtium, corn
Strawberries	Bush bean, spinach, borage, lettuce, thyme (as a border)
Sunflower	Cucumbers
Tomato	Chives, onion, parsley, asparagus, marigold, nasturtium, carrot
Turnip	Peas

Q. Are there certain plants that do not work well together?

A. A few. Among them: Fennel—Do not plant this near tomatoes or bush beans. Tomatoes—Do not plant next to kohlrabi.

Q. What is the best time of year to plant vegetables?

A. This depends on the area and which vegetables are being planted. The following chart will serve as a guide to planting many of the most popular vegetables.

Vegetable Planting Table and Requirements

Planting Directions

Vegetable	Time to plant	Distance between plants (inches)	Distance between rows (inches)	Row-feet amount of seed per 100	Yield per 100 row-feet
asparagus	spring	18	48-60	Roots	12-24 lb.
beans, bush	1-8 weeks after last spring frost	4-6	18-24	1 lb.	50 lb.
lima	2-6 weeks after last spring frost	6-10	24	1 lb.	60-75 lb.
beets, early	2-4 weeks before last spring frost	3	12-18	2 oz.	100 lb.
late	6-8 weeks before first fall freeze	3	12-18	2 oz.	100 lb.
broccoli	4-6 weeks before last spring frost	18-24	24-30	plants	50 lb.
cabbage, early	4-6 weeks before last spring frost	15-18	24-30	plants	100 lb.
late	3 months before first fall freeze	24-30	24-30	1 pkt.	175 lb.
carrots, early	2-4 weeks before last spring frost	3	12-18	1 oz.	100 lb.
late	10 weeks before first fall freeze	3	12-18	1 oz.	150 lb.
cauliflower, early	2-4 weeks before last spring frost	18-24	24-30	plants	45 heads
late	3½ months before first fall freeze	18-24	24-30	1 pkt.	45 heads
corn, early	on frost-free date	12-18	24-36	4 oz.	100 ears
late	10 weeks before first fall frost	12-18	24-36	4 oz.	100 ears
cucumbers	1 week after last spring frost	36-60	36-60	½ oz.	150 lb.

	When to plant				
eggplant	1 week after last spring frost	24-30	24-30	plants	125 fruit
lettuce, head	4-6 weeks before last spring frost	6-12	12-18	½ oz.	50 lb.
leaf	6 weeks before first fall freeze	6-12	12-18	½ oz.	50 lb.
muskmelons	1-2 weeks after last spring frost	48-72	48-72	½ oz.	50 fruit
onions	4-6 weeks before last spring frost	2-3	12-18	300 pl.	75-100 lb.
parsnips	2-4 weeks before last spring frost	3-6	18-24	¼ oz.	100 lb.
parsley	2-4 weeks before last spring frost	3-6	12-18	¼ oz.	50 lb.
peas	4-6 weeks before last spring frost	1-3	18-36	1 lb.	40 lb.
potatoes, white	4-6 weeks before last spring frost	12-15	24-30	6-10 lb.	75 lb.
sweet	1-2 weeks after last spring frost	12-18	30-48	plants	100 lb.
radishes	2-4 weeks before last spring frost	1	12-18	1 oz.	1,200
rutabaga	3 months before first fall freeze	6-10	18-24	¼ oz.	150 lb.
soybeans	on frost-free date	6-10	24	½ lb.	50 lb.
spinach	4-6 weeks before last spring frost	2-6	15-24	1 oz.	50 lb.
squash, summer	on frost-free date	36-80	36-80	½ oz.	100 fruits
winter	1-2 weeks after last spring frost	48-120	60-120	½ oz.	100 fruits
strawberries	spring	12-18	36	plants	varies
tomatoes	on frost-free date	24-48	24-48	plants	200 lb.
turnips	4-6 weeks before last spring frost	3	21-18	½ oz.	100 lb.

Vegetables—Problems

II

Q. Why do the leaves on tomato plants sometimes curl up? Is this a sign of some disease?

A. Not usually. More than likely it is what is termed "physiological leaf roll," and is often noticeable after heavy watering or deep cultivation. Wilting generally precedes leaf rolling, but as new roots form, it will disappear. Tomato plant leaves roll upward and in toward the main vein. The exact cause is unknown, although some agronomists believe the condition stems from an accumulation of carbohydrates in the plants. Heavy pruning or removal of suckers sometimes favors this problem.

Q. What causes the blossoms to drop off tomato plants without setting fruit?

A. Tomato and pepper flowers often fail to set fruit for several reasons. Blossoms may drop because of excessive nitrogen fertilization, lack of polination, use of hormone-type weed-killers, or because of hot, drying winds. A balanced plant-feeding program, applying natural fertilizers that are gradually available—not forced as chemicals are—prevents too much stem and foliage growth at the cost of fruiting. Weed-killers used on lawns will volatize and drift over garden areas, causing plants to drop their blossoms and become stunted. Mulches, windbreaks and irrigation can help overcome the hot-weather factor. And honeybees, which carry pollen from male to female blossoms, are killed by insecticide spraying. Using non-poisonous pest control methods avoids loss of bees and reduced fruit set.

Q. What causes the cracking of tomato fruits?

A. Either a deficiency of minerals in the soil or too much water in the soil.

Q. What conditions favor blossom end rot of tomato fruits?

A. This rot is closely associated with a deficiency of water. It can be prevented by giving the tomato plants a good soaking with water at intervals until drought conditions are over.

Q. Is it absolutely essential that tomato plants be staked?

A. No. But unstaked tomato plants should be mulched with loose coarse materials to keep the fruits off the ground.

Q. Are there any general rules for tomato growers to follow?

A. Yes. In addition to certain specific measures for specific troubles, there are several general practices that all tomato growers should follow to keep losses from insects and diseases at a minimum.

Plant tomatoes in clean soil.

See that your seedbeds receive the proper amount of ventilation. This practice, together with the proper spacing and watering of the seedlings can do much to prevent disease in young plants.

Use plenty of properly-composted humus to build up the soil and promote the growth and general health of the plants. Plants grown in soil that has a liberal amount of organic matter will be more resistant to insects and disease. If the soil is not organically rich and diseased plants do appear, remember to pull them up and destroy them. Do not place these plants on the compost pile, or the disease or insects that have killed them may return to the soil.

Growing tomatoes on a large scale necessitates the rotation of crops. Once every three years is often enough to grow tomatoes on the same land. It is not a good practice to grow tomatoes in rotation with potatoes, eggplants, okra or peppers, as these crops are susceptible to the same diseases.

Don't walk through or cultivate the field when the

plants are wet. This will eliminate the unnecessary spread of such diseases as gray leaf spot, early blight, late blight and Septoria leaf spot.

Use resistant varieties wherever possible. The following varieties were reported resistant to a wide range of diseases: Step 348 Manapal, Indian River and Manalucie.

Q. What's the best method of getting well-formed carrot roots?

A. The soil in which carrots are to grow must be well prepared and enriched. It should be deep, mellow and friable. This should be free from lumps and stones which often force the roots into deformities and cause them to split. Nematodes are also a frequent cause of misshapen carrots. Large applications of humus may prevent nematode damage.

A good supply of humus from a well-made compost heap will do much to put the soil into condition. Exhibition specimens are sometimes grown by drilling holes in the ground and filling them with a mixture of equal parts of compost, leaf mold and sand. Apply finely ground limestone when soil tests show it is needed to correct excess acidity.

Deep cultivation frequently injures carrots because their feeder roots remain near the surface. Use a fine-shredded mulch to control weeds. Thin carefully when carrots are about half an inch in diameter so that remaining plants stand approximately two inches apart.

Q. What can be done to overcome crop loss to bean rust, other than using fungicide sprays?

A. Don't plant in soil where rust has infected a previous crop; choose varieties which are rust-tolerant (White Kentucky Wonder 191, U.S. No. 3 Kentucky Wonder, and Dade); avoid highly susceptible varieties (Blue Lake, McCaslan, Kentucky Wonder); and consider prevailing winds when choosing a planting site if there is a possibility of spores blowing in from an infected field. Since rust spores can be carried long distances, a long crop rotation is also advisable. The fungus lives over winter in old bean stems, but does not survive more than one year and is not seedborne.

Q. How can the bugs that plague squash, melons and cucumbers be controlled?

A. These and related crops are all victims of the same trio of main troublemakers—the squash bug, the striped cucumber beetle and the squash borer.

The squash bug, or stink bug, is especially fond of summer squash, laying reddish-brown eggs on the insides of the leaves. Pick off and crush these clusters before the gray nymphs hatch out. Older brown adults can also be hand-picked or trapped under boards set beside plants and turned over each morning to clear ambushed bugs. To deter them from plants, grow some radishes or strong-smelling nasturtiums nearby.

Striped cucumber beetles, which spread bacterial wilt besides doing other damage, appear as early as the sprouted vegetable seedlings. They are good reasons for sowing each hill thickly, then thinning to one sturdy specimen when plants are four to five inches high. A ring of radishes around each plant also helps repel the beetle, as does planting a few nasturtium seeds at the same time you sow a hill of squash. Another deterrent is a sprinkling of wood ashes or ground limestone about the base of the plant and on the foliage. Put it on when leaves are wet and repeat after rain.

The squash vine borer works inside the stem, often causing the main root to die. To combat this culprit, encourage formation of new roots by mounding earth over several leaf joints. Start vine crops as soon as possible after soil warms up. Well-established plants are better able to resist both the borer and squash bug, which appear a little later than the striped beetle. In infested vines, slit the stem with a sharp knife where a small pile of "sawdust" shows the borer is present and dig it out. Cheesecloth tents or plastic netting covers may also help curb all three insects.

Q. What should be done to avoid having tomato plants that show poor growth, become very yellowish in leaf color, and have hard, purple stems?

A. These symptoms indicate nitrogen deficiency. Improvement will follow applying organic matter rich in nitrogen, such as manure, cottonseed meal, dried blood and leguminous material (clover, alfalfa). If the ground is not needed for planting this year, perhaps you should grow a cover crop of legumes. After the crop is well established, apply a liberal covering of manure, finely ground raw limestone and finely pulverized phosphate rock. Work the cover crop and other materials added into the surface layer of soil. This practice will increase the nitrogen content directly through the decomposition of the legume crop and indirectly through rapid increases in the biological activity of the soil organisms, resulting in greater nitrogen-fixation.

Fertilizing Vegetables
III

Q. What are the advantages of broadcasting fertilizers during preparation of vegetable seedbeds before planting?

A. Broadcasting fertilizers before or during seedbed preparation raises the general fertility level of the soil. Spread the fertilizer by hand or with a fertilizer spreader, but mix it well with the soil by disking or raking. It is a good idea to make a broadcast application annually and it may be supplemented by applying fertilizer along the row after plants have started to grow.

Q. Why do some people sow fertilizer in a narrow furrow along the seed row at the time of planting?

A. This practice puts the plant nutrients close to the roots and makes them readily accessible to the growing plant. The fertilizer should be placed slightly deeper than the seed, but never in contact with the seed. Placement about one-half to one inch to the side of the seed row and one inch deeper than the seed usually results in effective use of the fertilizer.

Q. Is it a good practice to place the fertilizer around the hill at the time of planting?

A. This is an efficient way to use fertilizers for widely spaced plants. Place the fertilizer in a circular band around the hill, two or three inches away from the seed and one inch deeper than the seed.

Q. How deep should fertilizer be placed when it is applied along the row after the plants have made some growth?

A. It should be placed in furrows about three inches deep. If the plants are small make the furrows about two inches to the side of the plants. If the plants are larger make the furrow as close to the roots as possible without cutting or breaking them. After the fertilizer is placed in the furrow fill it with soil and moisten the soil by sprinkling.

Q. Are there any rules about applying fertilizer in solution?

A. Fertilizers which dissolve readily in water can be applied in solutions. Avoid strong concentrations on the foliage or too close to the roots. Compost of manure water is recommended.

Q. Why is it that some people use manure and kitchen wastes in their vegetable garden and still obtain poor results?

A. That's not enough. While manure and plant refuse will help improve the soil structure and will add needed organic matter, these materials alone won't correct an imbalance of essential nutrients, especially in a poor soil. Minerals, in particular, are vital to strong plant growth; although manure and kitchen wastes contain a small amount of minerals, the quantity and availability to growing plants is limited. Along with these humus-building materials, some natural ground rock fertilizers—phosphate and potash rock or greensand—should be incorporated to supply these major mineral elements. A soil test will show which factors are low in the soil, and indicate the amounts required to bring it up to a good nutrient balance. Also important are the pH levels and nitrogen supply, which will similarly be shown by a soil test.

CHAPTER 10

LAWNS

The one thing a gardener cannot afford to ignore is his lawn. For one thing, it's too obvious to hide so he has to mow it, whether he wants to or not. He must also work to keep it green and weed-free, because a poorly kept lawn ruins the appearance of his whole property. There is a "right" time to plant a lawn, a "best" time to mow it and even a "proper" height for the mower to be set.

Grass likes fertilizer—the right kind at the right time— as much as any other growing thing. But be careful what you feed your lawn and how often you feed it. A lawn likes water too, but there is a happy medium between a lawn that is drying and one that is drowning. The right information on the best way to maintain a lawn can save a lot of work and a lot of money.

Q. I have been asked to decide on the basic type of lawn we will have around the new home we are planning to build. What are the basic types of lawns? How can I judge which would be best for my needs?

A. The home owner should know about the three basic types of lawns and be able to recognize each of them on sight. He must decide what kind of lawn he wants to grow on his property, and make sure that local conditions will permit him to do this successfully. Different kinds and combinations of grasses make different lawns and specific soils are obviously better suited to one type of lawn than another. The home owner should know what lawn type he wants before he starts ordering seed.

1. *The All-Purpose Lawn* is a compromise lawn

121

An attractive lawn is the result of good soil, right seed and natural treatment.

which does equally well in sun and in moderate shade. It asks few favors and is usually the best choice.

2. *The Elegant Fine-Bladed Turf* is the show-place kind of lawn and everybody, for obvious reasons, can't have one. It calls for extra care and the grasses, mostly the bents, are shallow-rooted and discourage easily.

3. *The Extra-Sturdy Working Lawn* for the growing family with plenty of children and pets who want to romp and play all during the hot summer when the grass is most vulnerable. Remember no grass is indestructible, but the tough fescues will take considerable punishment— providing they get adequate care.

Q. Why is it necessary to test soil before planting a lawn?

A. Few soils have all the right nutrients in just the right amounts for ideal lawn growth. Some may be deficient in one element; others may lack two or more. Even within a single lawn plot, different sections often have varying levels of the same nutrients. Then, too, each time your lawn is grown, changes take place, depending on what practices are followed. Just standing idle will alter a soil's makeup from year to year. The more a home owner finds out about his soil, the better he will be able to provide for his lawn. Soil testing is one very practical way to learn more about the makeup of any soil.

Q. How do I go about preparing a soil sample for testing?

A. Assemble the tools, making sure they and the containers to hold samples are clean and free of foreign matter. Be especially sure there are no residues or fragments of fertilizer, which can throw the test results completely off. The digging or sampling tool may be a spade, trowel, garden dibble, narrow shovel with straight sides, or a soil auger or probe. For gathering samples, use a large bucket or similar container. Coffee cans or small cardboard boxes which will hold about a pint of soil will serve for the final samples.

Soil samples can be taken at any time during the year weather conditions permit. The soil should be free of frost and fairly dry. If the lawn is quite large, taking a composite, or mixed, sample from several points will provide more information than a spot sample. (However, a combined sample from two distinctly different soil types would not be a suitable representative of either one. In a case like that, individual samples would be better.)

For a composite sample, start at one end of the plot and with the garden tool cut straight down about 6 to 8 inches and lift out a narrow slice; lay that to one side. Then take another thin slice, about $\frac{1}{2}$ inch, from the same section and put it in the bucket or collecting container. Shovel the first slice back in place. Now move about 6 feet in any direction and repeat the operation. Continue until the plot is given an over-all sampling. On a field of an acre or more, samples can be taken several rods apart.

Once you've covered the area to be tested, mix the samples thoroughly in the bucket and remove one pint of soil. Spread this "homogenized" sample where it can continue drying, and when completely dry and without lumps, place it in the sample container. Damp or wet soil will give false test readings.

Remember, soil tests can be no better than the samples tested. To reflect an accurate picture, they should be as representative of the area as possible; otherwise the story told by the test will not be true. A pint of soil—the size of the composite sample recommended—is about 1/150,000,000 part of an acre. A level teaspoonful—the amount actually tested—is about 1/100 part of a pint.

Q. What methods are available to the average home owner for testing soil?

A. The simplest to get and use is the home soil-test kit, available at many garden-supply stores. It contains materials and instructions for determining the approximate content of the 3 major soil nutrients— nitrogen (N), phosphorus (P) and potash (K)—and for establishing the pH standing.

Another choice is to send a sample of soil to a laboratory for testing. Farmers, rural homesteaders and city gardeners frequently get help and information through their county agent. Most state experiment stations charge a small fee for the soil test report, which in addition to an NPK and pH analysis may include indications of trace element levels and organic matter content. Stations tend to be slow in responding especially during planting and fertilizing periods.

Q. We are readying our lot for its first lawn. How deep must we plow and what should we guard against?

A. Quality grasses cannot thrive on an infertile, hard-packed, or rubbish-filled soil. A loose, porous structure is needed, one that air, water, and roots can penetrate. Grass can stand a moderate amount of competition, but cannot battle against many surface tree roots.

When you build a lawn on a new lot, start properly. That includes a decent burial for a contractor's quick-grow seeding. Spade, till, or plow and disk at least 8 inches deep. Remove all debris, plaster scraps and large stones.

Q. What must be done before replanting an old lawn?

A. If it is tired, hard-packed, and sparse, prepare an old lawn in the same way as you would a new lot. Skip areas of exposed tree roots and plan on a ground cover. Leave areas next to house walls for shrubs or flowers.

Q. What is the most desirable rate of slope in grading?

A. Most lots have some fixed grade points. These points are the house foundation, sidewalks, driveways, terraces and established trees. In grading, both rough and finish, the problem is to spread the soil evenly so that the changes in elevation between fixed points are gradual. In general, the best grades slope gently away from the buildings in all directions. A slope of 6 to 12 inches per 100 feet is the most desirable.

Q. How should the seedbed for a lawn be prepared?

A. Much of the success of a new lawn depends on how well you prepare the seedbed. Therefore, allot ample time and care to this phase of lawn establishment. The seedbed preparation establishes the final grade and makes a firm, smooth soil surface ready for seeding.

Slope the subsoil of lawn seedbed away from the house, leveling out uneven places with hand rake. (CREDIT USDA)

After the bulldozer has spread the topsoil, or if the area was worked up by plowing or rotary-tilling, then rake the area. Raking should remove stones and other debris as well as smooth out high and low spots.

Next, add the desired organic material, fertilizer and limestone. A second raking will thoroughly mix these materials in the soil.

It is usually desirable to roll the seedbed before seeding. However, you may seed directly in this loosened soil if it is firm under the first inch or two. The depth of your footprints will reveal the firmness of the soil.

If you roll the soil before seeding, rake the area over lightly to loosen the surface.

After sowing the seed, rake the surface again, lightly, to cover most of the seed and roll it again.

Spread lime, if needed, and phosphate fertilizer over subsoil before covering with top soil. (CREDIT USDA)

This compact seedbed will offer the best condition for germination of the seeds and development of the young grass after germination. If the seedbed is too loose and deep, moisture may not be sufficient for germination and early growth of the grass.

Q. Do you have any tips on seed buying?

A. There are over 1,500 species of grasses in this country, but only a few are suitable for lawns. Some are coarse and quick-growing, like ryegrass. Some die out in hot weather, like Canadian bluegrass. Some stand up well and spread rapidly, like Merion bluegrass and the bents. Some grow in the shade, like the fescues, and some in full sun, like Kentucky bluegrass. Some must be started from stolons, especially in hot, dry climates. The inclusion of clover is a matter of choice.

Buy the best quality lawn seed for your section of

the country, sunny or shady as appropriate. Make sure that the germination rate of the seed is good, and that the chaff and weed-seed content of the mix is low. Use the seed the same year you buy it. Don't buy so-called bargain seed, containing a high percentage of ryegrass, timothy, red top, or Canadian bluegrass. Avoid unproven, wonder-promising lawn seeds. Don't try to produce a good lawn by mowing field grass and don't expect even the best seed to grow on rock-hard infertile soil.

Q. What is the purpose of using a seed mixture rather than a single kind of good grass seed in planting a lawn?

A. A mixture has a wider adaptation to soil and climate conditions than a single grass. Seed mixtures also have more tolerance to diseases, insects and weeds. A third reason for using a mixture is that it makes a good sod sooner after seeding. This is particularly true with a spring seeding.

Q. When is the best time to seed a new lawn?

A. From late August to late October is the best time for seeding a new lawn, especially if you live in the northern part of the country. The warm soil, usually adequate rainfall, and cool nights combine to establish an ideal growing condition for lawn grasses. In addition to this, fall-sown grasses root more deeply and are unlikely to be choked out by weeds, as fall is a dormant period for weeds.

Q. What grass should be used in problem areas such as playgrounds, banks and traffic areas?

A. There is an increasing need for specialized turf areas, such as athletic fields, playgrounds, public building grounds, parkways, fairways, cemetery and highway rights-of-way.

The factors to consider in grass selection should include degree of traffic, slope, light intensity, nature of soil, desired appearance, clipping height and exposure. The characteristics of each grass must be understood to make the proper choice. For example, tall fescue for a football field is a mistake because it requires a high clip. But tall fescue, as well as Common Bermuda grass, is well adapted to heavy traffic areas and playgrounds. Common Bermuda grass is the best for football and baseball fields.

Q. Is there any kind of grass that is especially recommended for problem lawns?

A. Yes. If you're looking out on a sad substitute for a lawn, plant zoysia.

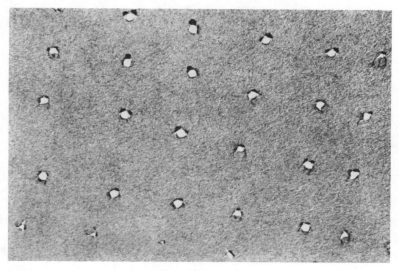

Fit plugs into prepared holes about one foot apart.

Q. What are the advantages of a zoysia lawn?

A. 1. Zoysia grass lawns are permanent, lasting many years.

2. Zoysia grasses make a tough, wear-resistant, cushioney turf.

3. They grow slowly . . . the average home owner may be satisfied with the appearance maintained by mowing 10 days to 2 weeks apart.

4. Zoysias thrive on high temperatures of summer.

129

A power mower must be used in cutting Zoysia.

5. They grow on a wide range of soils, from sandy to heavy loams if there is good subsoil drainage.
6. They make reasonable growth and cover with little fertilizer, but will have a better color if well fertilized.
7. The density of a mature turf will crowd out weeds.
8. Zoysias are practically free from disease and insect pests.
9. They grow well in moderate shade or full sun.
10. Zoysias are salt and chlorine tolerant. They may be used around swimming pools.

Q. Are there any disadvantages to using zoysia grass?
A. Yes, it has several drawbacks.
1. Zoysia grass will lose some green color in winter. The length of this period of discoloration will vary from a few days to a month or more, depending on location and weather.
2. Ten to fifteen months are required to produce

sod in various climatic areas of California. Weeds must be controlled during this time.

3. Zoysias must be planted by sprigs or plugs or turf because seed is unreliable, slow to germinate and variable.

4. A power mower is required for easy cutting.

5. The density of a mature turf prevents successful use of cool-season companion grasses for winter turf in cold areas where off-color period is long.

Q. Are there any rules to follow when mowing a lawn?

A. Yes. There are three rules for sound mowing practice:

1. Mow high, even as high as three inches or more if your mower permits.

2. Mow regularly, as often as grass growth requires.

3. Never cut more than one-third of the total length of the grass blade at one cutting.

Q. Is it all right to leave grass clippings on the lawn or must they be raked?

A. If your grass is cut once a week, it is good to leave the clippings right on the lawn so that they may become natural humus. If the growth is particularly heavy it is best to rake; a heavy crop of clipping leads to wads of yellow hay.

Q. I have a large lawn that takes so long to cut that I have little chance to do any other gardening. What can be done to save on lawn mowing time?

A. 1. Provide for continuous mowing by avoiding sharp corners and angles around plant beds and building corners.

2. Use a section of flush paving around lawn obstructions such as fire hydrants, light poles, sign posts and sewer vents. This can eliminate hand trimming and speed-up power mowing.

3. Avoid impossible-to-mow situations. Use low maintenance groundcover on steep slopes and bumpy or rough areas.

Metal edging cuts down on trimming chores.

Q. Do you know any tricks that will ease trimming chores?

A. 1. Use concrete, brick or stone mowing strips against buildings, walls and under fencing to eliminate hand trimming.

    2. Keep lawn areas flush with paved surfaces such

as walks and terraces to avoid unnecessary trimming and provide easier movement of maintenance vehicles.

3. Eliminate hand trimming around trees by using grass barriers or metal edgings. This will also reduce tree damage from mowers.

Q. How often should a lawn be watered?

A. Many gardeners make the mistake of watering their lawns every day it doesn't rain. Lawns that are watered every day, or even every few days, lose their ability to send their roots deep into the ground in search of moisture and food. They develop shallow root systems and as a result will brown easily. A thorough soaking once every five days or a week is sufficient to keep a lawn green and healthy.

Q. When is the best time to fertilize grass?

A. That depends on the type of grass you grow. Cool season grasses should not be fed when they are going into their semi-dormant summer period. Nitrogen supplied in late spring will benefit the weeds, not the grass. Warm-season grasses, on the other hand, should have their heaviest feeding at that time. When the new shoots of Bermuda grass or zoysias appear, a good top-dressing with a nitrogenous fertilizer will help them to recover from winter.

Q. Do all lawns need regular applications of lime?

A. No, they don't. Lime requirements depend on several factors: where you live, the type of soil under your lawn and the sort of fertilizing program you practice. Where soils are naturally alkaline, they should receive no lime. Where organic matter and humus content is high, it acts as a pH (acidity-alkalinity) buffer and makes liming necessary less frequently. Highly acid fertilizers, such as many sulfates and other chemical nitrogen types, increase lime demands. Organic forms of fertilizer are more gradually available, tend to overacidify much less.

Many home owners damage their lawns by "feeding" them lime. Lime applications can improve growth and appearance of lawn grasses, but when not needed they can cause trouble. The best way to get an accurate test to determine if lawns need lime is to take a sample of the soil to your county extension office or to make a careful soil test yourself.

Q. How should compost be prepared so it won't detract from the appearance of a lawn?

A. Since ordinary compost is much too lumpy for use on lawns, it should be thoroughly ground up. A shredder or a rotary mower can be used for this. If this equipment is not available place the compost on a wide board and work it back and forth with a hoe until it is shredded very finely. Then work it through a sieve (about one-half inch mesh). Apply the coarser residue to the vegetable patch or work it through the sieve again.

Q. Where can ground covers be used to best advantage?

A. Does your lawn have bare spots that even Houdini couldn't grow grass on? There are ground covers that will succeed where all else fails. Problem areas that also benefit from ground covers are flagstone steps, lawn and garden borders, odd angle corners and under and around permanent garden furniture —all places where it's difficult to cut grass.

Q. Do ground cover plants need special care?

A. Almost all the common ground cover plants are as vigorous as weeds. Dig to a depth of at least 8 inches, removing all weeds and foreign material. Mix in a 2 to 3 inch layer of organic material such as compost, peat moss, rotted manure or aged sawdust. Soak well a few hours before planting.

Once established, ground covers need no coddling. But while they are getting started they should get some care. Water them occasionally during the first season. Topdressing with compost, humus, or

Most ground covers can be planted three to eight inches apart.

well-rotted manure will supply the nitrogen needed for dense growth.

Q. The initial cost of ground covers seems rather high. Can they be planted a good distance apart to economize?

A. Some gardeners make a practice of setting plants a good distance apart, thus using a smaller number of plants. However, they must be prepared to battle weeds in the longer interval it takes the plants to cover. You can buy seed, seedlings, or rooted cuttings from nurseries in order to produce the ground cover more cheaply at home. It's best to start these in your "home nursery" or in pots, carefully shading and watering them before setting them out.

Some thrifty gardeners search out wild plants from nearby fields and woods and transplant these.

The ideal method is to set small plants fairly close

together, depending on the rate of growth and size of the plants at planting time. Three to eight inches apart is correct for most ground covers, while some vines can be effectively planted as far as 36 inches apart.

Q. Ground covers sound too good to be true, don't they have any drawbacks?

A. You wouldn't expect grass to grow on a windswept or rainwashed slope; neither will most ground covers. None of the so-called ground covers can cover the soil as closely as a good grass turf. They are not recommended for any area where there is traffic—none of them will stand up to wear.

Q. What is thatch?

A. Thatch may be defined as a tightly intermingled layer of partially decomposed leaves, stems and roots of grasses which develops beneath the actively growing green vegetation of the soil surface.

Q. Is thatch a serious problem in a lawn?

A. Thatch decreases the vigor of turfgrasses by restricting the movement of water, air, plant nutrients and pesticides into the soil. During wet periods it acts as a sponge and holds excessive amounts of water. Many turfgrass disease organisms may be harbored in thatch accumulations.

Q. How can thatch be controlled?

A. It can be prevented by liming, fertilizing, mechanical thatching and aerating. Clipping removal should also be given consideration.

In a curative program where large amounts of thatch have accumulated, it is unwise to attempt to remove all of it in one treatment. Thatch removal is done mechanically and the machines fall into three basic groups: those machines having revolving swinging knives, those having revolving stationary knives, and those having revolving spring tines. If considerable thatch accumulation has occurred and deep penetration is required for removal, a machine having solid or swinging knives should be used.

Q. What conditions make lawns susceptible to disease?

A. Disease is encouraged by:
1. poor soil drainage;
2. excess moisture;
3. poor circulation of air because of surrounding trees, shrubs, or buildings;
4. incorrect mowing;
5. stimulation of grass with fertilizer during the summer;
6. strong soil acidity.

Q. Why do experienced gardeners place such stress on proper drainage and moisture control?

A. Correcting poor soil drainage and maintaining adequate soil aeration builds your lawn, because they permit stronger root and top growth. On the other hand, a soggy soil that is water-soaked makes an ideal environment for a sick lawn, because the disease organisms need an abundance of moisture for the early stage of spore development and infection of the plant.

Watering late in the evening is responsible for more lawn disease than perhaps any other bad practice. This is because the grass remains wet through the night, thus directly encouraging mold and fungus growth.

Q. Can you describe powdery mildew and tell what causes it?

A. Powdery mildew shows as isolated wefts of fine, gray-white, cobwebby growth mainly on the upper surface of the leaf blade. The growth becomes more dense, and the leaves appear to have been dusted with flour or lime. Infected leaves usually turn yellow and wither. It is found more commonly on bluegrass (especially Merion) in the spring and the fall when nights are cool.

Mildew is caused by a fungus which grows on the surface of the leaf. Sucker-like structures grow into the outer leaf cells from which the fungus obtains its nourishment. Mildew is most severe in shaded areas or where air circulation is poor.

Q. How can I prevent or eliminate powdery mildew?

A. To control powdery mildew, keep the lawn growing vigorously by fertilizing and maintaining adequate moisture in the soil. Improve air circulation to remove humidity pockets by pruning trees and shrubs where possible. Merion bluegrass is more susceptible than common Kentucky bluegrass.

Q. My lawn is littered with mushrooms. Do they do any damage to the grass? Is there any way to get rid of them for good?

A. Mushrooms, commonly called toadstools, often appear in lawns during rainy spells in the summer. Mushrooms are the aboveground growth of certain fungi which grow on decaying vegetable matter in the soil. In lawns this organic matter frequently consists of buried stumps or tree roots. Mushrooms are chiefly a nuisance as they do no harm to the grass and are best removed by raking or sweeping. No material will kill the fungus without injury to the grass. When toadstools or mushrooms appear, test your soil to check its pH. Then top-dress with enough natural ground limestone—preferably dolomite, which supplies magnesium as well—to bring it up to a desired level. Pulverized oyster shells, wood ashes, and other calcium-rich wastes are also usable. Best pH for healthy lawn growth is slightly acid—about 6.5—or just under neutral (7.0). To raise a soil one full unit—say from 5.5 to 6.5—apply approximately 50 pounds of lime per 1,000 square feet, 10 to 15 pounds less for light, sandy soils, more for heavier loams or clay types. If a large amount is needed you can apply half in early spring, the other half in fall.

Q. Is slime mold, the grayish-white patches that form on lawns, harmful to grass or soil if untreated? What

138

can be done to get rid of it?

A. Slime mold—small white, gray or yellow slimy round masses growing over grass blades—shouldn't cause alarm or trouble. It's harmless and if left alone soon disappears. While most common in the spring, slime molds may occur in midsummer or fall following heavy rains. The mold inhabits the soil, feeding on decaying organic matter; in humid weather the mold grows out of the soil onto anything available for support and produces its spore masses. The colored patches dry to form bluish-gray or white powdery structures that disintegrate when crushed between the fingers and easily rub off the grass blades. Slime mold can be removed by using a forceful spray from the garden hose or by raking the grass.

Q. Is there anything I can do to get rid of the powdery yellowish coating on my grass?

A. Rust fungi attack many lawn grasses, but are more serious on Merion Kentucky bluegrass than on other varieties. Symptoms are yellow-orange or red-brown powdery pustules that develop on leaves and stems. If a cloth is rubbed across the affected leaves, the rust-colored spores adhere to the cloth and produce a yellowish or orange stain. Rust usually occurs in late summer. Heavy dew favors rust development.

Damage from rust is less severe if Merion Kentucky bluegrass is mixed with common Kentucky bluegrass or with red fescue. The best control is to keep plants growing rapidly by fertilization and irrigation. A well-managed program of lawn care that includes the proper mowing, organic fertilizing and watering will go a long way in the control of rust.

Q. Do ants and anthills do more damage than good to lawns?

A. Ants build nests in the ground and usually form mounds around the openings. The anthills may

139

smother the grass under them as well as destroy the grass roots in the immediate area of the anthill. Although anthills annoy many people, especially in lawns where the grass is kept short, actually, lawns could not do better than to contain large numbers of these natural aerators. If the hills and castings are unsightly, allow the grass to grow a little longer, so that they will be hidden.

Do not use cyanogas or other chemicals for exterminating ants, as this will impair the soil. If ants are a problem, try sand, bone meal, or powdered charcoal. Once land becomes rich organically, the ant problem will take care of itself. Organic matter in the soil increases the moisture content and makes conditions less favorable for the ants.

Q. Chinch bugs wreck the looks of a lawn like nothing else can. Is there a way to eliminate them without poison sprays?

A. Hairy chinch bugs are serious pests of lawns. They are likely to be more serious in lawns containing bentgrass; however, bluegrass is also attacked. Damage to lawns by chinch bugs is caused by the young bugs or nymphs, which are about $\frac{1}{4}$ inch long, black with a white spot on their back between their wings.

Chinch bug infested lawns may have many large irregular dead patches. The bugs will be found within a circle of grass which has turned yellow around these dead spots.

To control this pest, seed lawn in soil made up of $\frac{1}{3}$ crushed rock and $\frac{1}{3}$ compost. Shade trees planted on the lawn may discourage the insect also.

Q. How do I keep my lawn weed-free without resorting to the commercial weed-killers that might kill everything else too?

A. 1. Fertile soil, especially soil well conditioned with organic matter, is not naturally conducive to weed growth.

2. Prevent seed production in nearby fields. Weed

seeds don't originate only from seed you sow. Weeds can be borne by wind, water, or even carried on clothing of people walking from nearby weed infested areas.

3. Hand pulling. Maybe it's a bit more work, but it's one of the most effective weed preventives known.
4. Mowing lawns properly is an effective method of keeping weeds in check because it allows the grass to crowd out the weeds.
5. Burning. Often noxious weeds which have begun to mature can be burned on the spot. Several small portable weed burners are available.
6. Fertilize organically.
7. Use clean seeds. Most often cheap seed is the most expensive you can buy because it is high in noxious weeds.

CHAPTER 11

INSECTS

All bugs aren't bad bugs. In fact, about nine out of every ten different species of insects in this country are beneficial to the garden. The important thing is to learn how to control the destructive pests without annihilating the helpful creatures at the same time.

The first step is to learn how to distinguish one insect from the other—which insects are to be discouraged and which are to be courted. Next you need to learn the easy organic ways to rid your garden of the real pests.

Insect control does not have to mean lethal pesticides.

Good and Bad Bugs

I

Q. Are there any basic rules to remember in the war against the garden's ever-present insects?

A. 1. You can't kill all the bugs. Many of the millions of insects in our country have shown surprising resistance to all kinds of environmental abuse.

2. All insects aren't bad. Ninety per cent of the species of insects in the U.S. are "friendly" or beneficial to the garden. Blankets of lethal spray and insecticides kill both the desirable and the destructive insects.

3. Soil comes first in controlling insects. Because insects have a craving for plants that are nutritionally deficient, you can offer the *best* protection by providing an organically fertile soil.

4. Discard the "kill at any cost" philosophy. Bugs are part of nature's scheme of things.

Q. What is the safest, most effective way to get rid of barn flies?

A. Your veterinarian will supply you with a rotenone or pyrethrum dust which may be used to dust animals. These are plant extractives and if obtained for use on animals must not contain any poison material.

Hanging fly ribbons, available from grocery and hardware stores, at various places in the barn will collect the flies. Make sure the ribbons are hung so that the animals cannot pull them down. Each ribbon contains a metal tack.

Hanging tapes or ropes which contain a poison deadly to the fly can be purchased and tacked on ceilings and over windows.

Experiments have proved that leaves from the cucurbit family (pumpkin and squash) crushed and rubbed on animals, repel flies. So will mint, marigolds and calendulas planted close to the barn.

Of course all manure must be removed daily, along with all trash, and the premises kept as clean as possible, for best results.

Q. What are some of the best methods for controlling insects without sprays?

A. 1. Be sure that you keep your plants healthy. Plants attacked by insects are often nutritionally unbalanced.

2. Use companion plantings. The garlic bulb is a favorite for this purpose. Other effective companion plantings are herbs, onions, mint, marigolds, pansies and nasturtiums.

3. Helpful insects like the ladybug and praying mantis make effective inroads by controlling insect villains.

4. There are also many biological controls on the market.

Q. What are some specific alternatives to insecticides in combating individual pests?

ANTS

A. Keep ants away by banding plants with tangle-foot. This prevents insects from climbing up the stalk. In home gardens, steamed bone meal has been found to discourage ants. If ants persist, try a pepper spray. Shellac the exterior of ruined bark or wood in nearby trees to take away their favorite habitats. Tansy has been found to discourage ants. Plant it at the back door, or around house foundations to keep both ants and flies from the house. The dried leaves of tansy sprinkled about indoors—in cellar or attic—make a harmless "insecticide."

APHIDS

Enrich your soil organically, as the aphid detests plants grown in organically rich soil. Grow nasturtiums, which repel aphids, between your vegetable rows and around fruit trees. Or make a trap of a small pan painted bright yellow and filled with detergent water. The aphids will become attracted to the bright yellow color and alight on the water surface, trapping themselves. You can trick aphids by placing some shiny aluminum foil around your plants so that it reflects the brilliance and heat of the sun. Aphids shy away from foil-mulched plants. Finally, encourage the aphid's natural enemy, the ladybug, who eats many times her weight in aphids.

BAGWORM

Bagworms are easily removed by hand. Black light traps are also effective for catching the worms in their moth stage. Trichogramma are a natural and effective enemy.

BEAN BEETLE

Encourage beneficial praying mantises, which have been most effective in bean beetle control. Also, plant your heaviest crop of beans for canning and freezing early, because those plants are freer of the pest than late season ones. Don't forget to use interplanting techniques with potatoes, nasturtiums, cloves and garlic. Some gardeners have used a mixture of crushed turnips and corn oil to foil the beetle.

CABBAGE MAGGOT

Create a strong alkaline area around the plants to deter the maggot by placing a heaping tablespoon of wood ashes around each plant stem, mixing some soil around with the ashes and setting the plants in firmly. Protective canopies of polyethylene netting also prevent infestation by keeping insects from laying their eggs in the young plants. In addition, practice general insect control measures such as the use of sanitation, rotation and improvement of the soil condition. Interplanting might be able to beat the cabbage worm. Surround your cabbage by cole plants such as tomatoes and sage which are shunned by the cabbage butterfly, the parent of the green worm. Further interplanting techniques include the use of tansy, rosemary, sage, nasturtium, catnip and hyssop. Friendly insects like trichogramma are available from commercial sources. Avoid using poison sprays which will kill these and most helpful insects. In addition, homemade, non-toxic sprays such as pepper sprays, sour milk sprays and salt mixtures have been found effective.

CATERPILLAR, TENT

Perhaps the best control for caterpillar is the use of *Baccillus thuringiensis*. This non-harmful bacterial is eaten by the insects who become paralyzed and die in about 24 hours. Other control measures include the use of sticky bands so that the female worms, who are unable to fly, will not be able to crawl up trees and lay eggs there. You may wish to place burlap or shaggy bands around the trees to attract the caterpillars and trap them. Those caught can be destroyed daily. Light traps are also effective. Encourage praying mantises and birds. Trichogramma is effective against the tent caterpillar as well as the gypsy moth and cankerworm. Remove and destroy the brown egg masses from the branches of any wild cherry trees in vacant lots and other areas around your home. Every time a single egg mass is destroyed, the potential threat of 200 to 300 more tent caterpillars next spring is gone.

CHINCH BUG

This little black sucking insect can cause large brown patches in your lawn and all but destroy your sweet corn. Chinch bugs thrive on nitrogen-deficient plants, so heavy applications of compost will make your plants resistant and avoid much of the trouble. If these bugs are present in your lawn, remove the soil from the spot and replace it with one-third crushed rock, one-third sharp builder's sand, and one-third compost. If they show up in the corn patch, plant soybeans as a companion crop to shade the bases of the corn plants, making them less desirable to the highly destructive chinch bug.

CODLING MOTH

Place a band of corrugated paper around the main branches and trunk of affected trees. When larva spin their cocoons inside the corrugations, they can be removed and burned. Eliminate loose bark from trunks and limbs where moths like to hibernate. Trichogramma, tiny female insects, are helpful in biological control of moths. Also effective is a black-light-bulb trap which attracts and kills the moths during their summer sessions. One gardener suspends a container of molasses and water mixture in his trees to trap the moth. Ryania will discourage codling moth, and birds are effective natural controls. Woodpeckers consume more than 50 per cent of the codling moth larvae during the winter.

CORN BORER

Destroy overwintering borers by disposing of infested cornstalks. Plow or turn under the refuge or relegate it to the compost heap. Plant resistant or tolerant strains of corn—consult your county agent for the best hybrids available locally. Because moths lay their eggs on the earliest planted corn, it is generally advisable to plant as late as possible, staying within the normal growing period for your locality. Encourage parasites like the spotted lady beetle which eats the eggs of the borer on an average of almost 60 per day. Trichogramma is also a natural enemy of the corn borer.

CORN EARWORM

Fill a medicine dropper with clear mineral oil and apply it into the silk of the tip of each ear. Apply only after the silks have wilted and have begun to turn brown. Another easier control, but not as sure, is to cut the silk off close to the ear every four days.

CUCUMBER BEETLE

Heavy mulching is a time-tested control. For every bad infestation, spray by mixing a handful of wood ashes and a handful of lime in two gallons of water. Apply to both sides of leaves. Radishes, marigolds and nasturtiums offer interplanting protection.

CUTWORMS

Cutworms chew plants off at the ground level. They work at night and hide beneath soil or other shelter during the day. A simple device for preventing damage is to place a paper collar around the stem extending for some distance below and above the ground level. Some gardeners get the same effect from using tin cans, with "electroculture" benefits as an added plus. Toads and bantam hens are natural feeders of cutworms. Cultivate lightly around the base of the plant to dig up the culprit first. Keep down weeds and grasses on which the cutworm moth lays its eggs. Interplanting with onions has been found to be effective in many cases.

EARWIGS

Traps have been an effective control for earwigs. Bantam hens feast on the earwigs and do a good job of eliminating them from the home grounds. Set out shallow tins of water which will attract earwigs so they can be destroyed.

FLEA BEETLES

Clean culture, weed control and removal of crop remnants will help to prevent damage from flea beetles. Control weeds both in the garden and along the margins. Since flea beetles are sometimes driven away by shade, interplant susceptible crops near shade-giving ones. Tillage right after harvest makes the soil unattractive to egg-laying females and will assist in destroying eggs already laid.

GRASSHOPPERS

Virtually every kind of bird has a craving for grasshoppers. Some eat the eggs after scratching them from the ground. Construct birdhouses and otherwise attract birds to your garden if you experience difficulty from grasshoppers. Grasshoppers can be baited by using buckets or tubs of water and a light placed nearby. Because grasshoppers lay their eggs in soil not covered with plants, keep a good ground cover to prevent egg laying in the soil. Turn the soil in the spring to a depth of 8 inches so the eggs will not hatch. Eliminate any weeds around garden margins.

JAPANESE BEETLES

Perhaps the most important control organism is the "milky spore disease," a bacterial organism that creates a fatal disease in the grub. The disease is caused by a germ not harmful to man and is available commercially. Since the beetles are attracted to poorly nourished trees and plants, be certain your soil is enriched by the addition of plenty of organic matter. Remove prematurely ripening or diseased fruit, an attractive dish for the beetles. Eliminate weeds and other sources of infestation like poison ivy and wild fox grape. Some gardeners get effective results by interplanting larkspur.

LEAFHOPPERS

Leafhoppers seem to prefer open areas, so plant your crops near houses or in protected areas to avoid damage. It's also a good idea to enclose your garden plants in a cheesecloth or muslin supported on wooden frames. Pyrethrum is an effective non-poisonous control that can be dusted on top of the plants. Keep the neighborhood clean and raked so that insects will be exposed to the weather, particularly during the autumn. Avoid planting susceptible varieties.

MAGGOT

Use tar paper collars around the stems to prevent the flies from laying eggs on the plants. Place plants in irregular rows so that the maggot is not able to find them easily. In the case of onions, this random technique often offers increased protection to neighboring plants because the onion smell is repulsive to

149

many garden pests. Hot pepper spray is an easy and certain control.

MEALYBUG

Wash off plants with a strong stream of water or use fir tree oil. Denatured alcohol can be used on house plants and may be successful in light infestations. Cultivate or turn the soil for several weeks before planting to kill any grass or weeds which may be hosts. Also scatter bone meal to ward off fire ants which often carry individual mealybugs from plant to plant.

MOSQUITO

Mosquitoes may be controlled by draining stagnant bodies of water or by floating on them a thin film of oil. While this may be somewhat injurious to vegetation, it is not as dangerous as DDT or other poisons. Often rain barrels and other containers with water become mosquito breeding places. Eliminate those from the home ground. Perhaps the best control in your immediate area is to encourage birds like the purple martin, just one of which will eat 2,000 mosquitoes a day. The praying mantis is also a natural enemy. Agricultural experiment stations have had some success with using a garlic spray.

NEMATODE

The consistent use of compost will virtually eliminate nematodes. Avoid chemicals to exterminate them, as that will interfere with the proper functioning of beneficial soil organisms which tend to keep out all dangerous microbes and nematodes. Organic fertilization of infected plants induces the formation of roots and improves plant vigor, thus negating the harmful effects of nematodes feeding on the roots. The most practical answer to the nematode problem for the average gardener is to

build up the humus content of the soil and to interplant with marigolds, especially the French or
African varieties.

PEACH BORER

Protect peach trees by keeping the ground beneath them perfectly clean of grass, weeds and
mulch for at least a foot in all directions to discourage rodents and other animal pests. This also
enables birds to get to the young borers. Swab each
one of your peach trees with tanglefoot before planting. The substance will catch the moths or the
worms so that they cannot penetrate the material
or get into the bark. Planting garlic cloves close to
the tree trunk has been found effective against the
borer as has the trichogramma.

RED SPIDER MITES

These pests thrive in stagnate, very humid air
so try to give your plants good air circulation. Remove mites from plants by spraying forcibly with
plain water, being sure to hit the undersides of the
leaves. (Generally, spiders washed off plants do not
return.) A three per cent oil spray has also been
found effective. Ladybugs are the mites biggest
nemesis, so encourage their visit to your garden. An
onion spray has been found effective but do not use
poison sprays as they usually kill the enemies of
mites but do not kill mites themselves. Pyrethium
is a safe dust-type control and can be used both
indoors and out.

ROOT MAGGOT

Repel root maggots by applications of large
quantities of unleeched wood ashes or a mulch of
oak leaves if available. Be sure to locate your growing area away from members of the cabbage family
for at least three years. Maggots are particularly

attracted to radishes, and some gardeners plant them as a trap crop to be discarded later. If infestation is heavy, test the soil and feed it. Then grow a cover crop to be turned under.

SCALE

Best control is to spray infected trees early in the spring with a dormant oil emulsion spray. Ladybugs, available commercially, feast on scale insects and usually keep these pests under control.

SLUG AND SNAIL

Snails and slugs tend to be nocturnal. Take advantage of their nighttime habits by placing shingles, planks, boards or other similar material in the garden to serve as traps. Each morning destroy those which have hidden away there for the day. The bodies of snails and slugs are soft and highly sensitive to sharp objects such as sand and soil and dry and slightly corrosive substances such as slaked lime and wood ashes. A narrow border of sharp sand or cinders around a bed or border will serve as an effective barrier against them as will a sprinkling of slaked lime or wood ashes. Many gardeners have found that setting out saucers of beer, sunk to ground level, attracts slugs by the droves so they can easily be destroyed.

SOWBUG

The best control for sowbugs is prevention. Look for and eliminate hiding places in and around the home garden area. Make certain that logs, boards and other damp places are eliminated. Frogs and poultry like to feast on sowbugs, so if you're lucky enough to have some around, turn them lose on this garden villain. If not, discourage them with a light sprinkling of lime.

SQUASH BUG

Squash bugs can be repelled from squash and other susceptible plants by growing radishes, nasturtiums and marigolds nearby. Hand picking of either the eggs, nymph or adult stage is effective in a small garden. Because the squash bug likes the damp moist protected areas, he often hibernates under piles of boards or in buildings. By placing boards on the soil around your plants you might be able to trap him and easily destroy the bugs every morning.

TOMATO HORNWORM

Tomato hornworms may be hand picked and killed by depositing them in a small can of kerosene. On a larger commercial scale, the tomato grower may obtain effective control from light traps, since the hornworm must pass through the moth stage in its life cycle. If the back of the hornworm is covered with a cluster of small white bodies, do not hand pick. Those are parasites which will kill the worm and live to prey on others. Trichogramma, praying mantis and a ground hot pepper dusting all prove good controls.

WHITE FLY

In the greenhouse or for indoor gardens generally, a planting of Peruvian ground cherry (*Nicandra physalodes*) is an effective white fly repellent. So is hanging fly ribbons on a stick. Outdoors, test for phosphorus deficiency in the soil. Ladybugs are fond of white flies, too. Many gardeners remove white flies from their garden area along with dandelion heads by using a vacuum cleaner to suck them up. In small greenhouses improved air circulation by exhaust fans is also helpful.

INSECTS

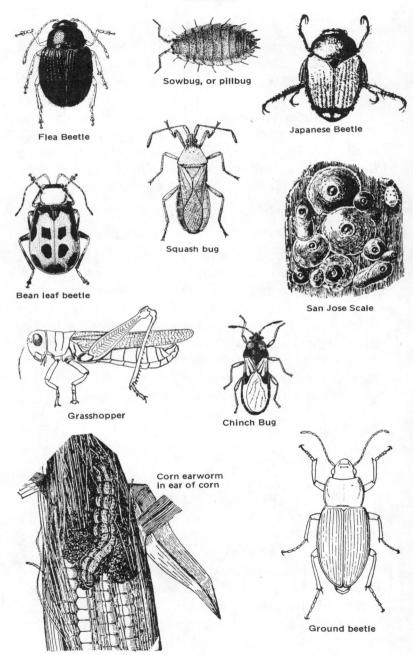

Flea Beetle

Sowbug, or pillbug

Japanese Beetle

Bean leaf beetle

Squash bug

San Jose Scale

Grasshopper

Chinch Bug

Corn earworm
in ear of corn

Ground beetle

INSECTS

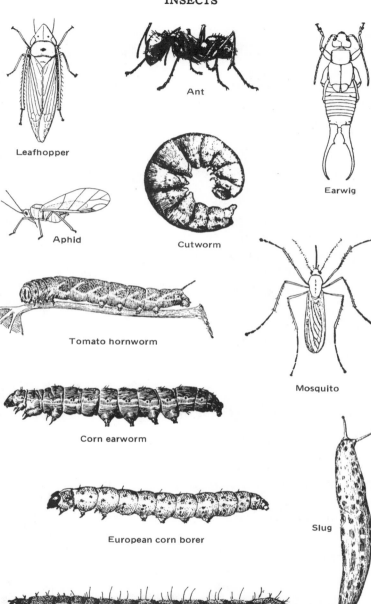

Leafhopper

Ant

Earwig

Aphid

Cutworm

Tomato hornworm

Mosquito

Corn earworm

European corn borer

Slug

Wireworm

155

WIREWORMS

Good drainage tends to reduce wireworm damage. Newly-broken sod land should not be used for the garden if other soil is available. If sod must be used, it should be thoroughly plowed or stirred once a week for several weeks in early spring. Stirring the soil exposes many of the insects and crushes others. Enriching soil with humus will also improve aeration and reduce wireworm attacks. Plant radishes and turnips as a trap crop.

All the insect controls suggested here have been edited and condensed from "The Organic Way to Plant Protection" published by Rodale Press, Emmaus, Pa. 18049. Should you desire more extensive control measures, we heartily recommend adding this book to your garden library.

Controlling Pests with Plants

II

Q. What are some of the plants found useful in controlling insect pests?

A. Cabbage butterflies and some other insects are said to be repelled by the scent from such plants as hemp, nasturtium, tomato, catnip, rosemary and sage planted near or among the cabbage and other vegetable plants. Japanese beetles are fond of geraniums, which are poisonous to them. The odorless African marigold has also proved a favorite of these beetles and diverts them from other plants. Among the many other plants which are helpful as insect controls are garlic, chives, thyme, tansy, rue, onions, zinnias, mint, basil, asters, cosmos and petunias.

Q. Is there any cultural way, besides biological control, to help eliminate onion maggots?

Basil is effective in controlling insects.

A. Adding sand to the top layer of planting rows will deter onion maggots. Radishes can also be used as an effective "trap" crop. If the infestation is heavy, planting cull onions at intervals near the seeded crop of field onions helps. The culls attract the egg-laying fly of the maggot and are then destroyed to keep the main crop clean.

157

Biological Control

III

Q. What is biological control of insects?

A. The practice of reducing the numbers of a pest by the use of natural agencies such as parasites, predators, and species-specific diseases is called biological control. It aims to achieve the most practical means of pest control by taking advantage of the fact that life in nature exists in a state of balance which is maintained by the competitive interaction of various forces. Man, however, through his diverse activities, frequently disturbs this natural balance —often with disastrous results to his own well-being.

For example, plant feeding insects are very seldom serious pests in their native environment where they have natural enemies that prey upon them. However, when such plant-feeding insects are transported to new areas and their natural enemies are left behind, they may become numerous and cause serious damage to crops. For this reason introduced species probably comprise two-thirds of the major insect pests of crops in the United States.

When such a change occurs in the natural balance of nature, it becomes the business of those concerned with biological control to re-establish a satisfactory balance through such control agencies as may be available regionally. Some of the methods used include importing natural insect enemies, sterilizing male insects, contaminating insects with disease, and breeding insects which will produce inferior progeny and releasing them in the area to be controlled.

Q. What are the advantages of biological control?

A. The advantages of the biological control of pests are so striking that every effort should be made to expand this means of pest control as rapidly as possible:

a) It is the most practical means of pest control. After the initial cost of the importation and distribution has been paid, there is no further expense.

b) The prevailing natural balance of potential pests in the area is not destroyed. In all agricultural areas there are many potential pests under natural control; the establishment of introduced enemies of a specific pest does not interfere with this natural balance of other and minor pests of the same crop.

c) There are no harmful effects to the health and well-being of plants protected by this method.

d) Pest resistance to biological control does not develop.

e) The biological method can in no way jeopardize the health of human beings or livestock.

Man & Beast

IV

Q. Besides herbs such as tansy, are there any plants that help to repel flies, especially around livestock?

A. Members of the cucurbit family, such as pumpkins and squash, make effective fly repellents. Nip leaves carefully from strong-growing vines. Crush them and rub on the backs and heads of cattle. For another fly (and flea) repellent, gather sprigs of mint leaves; hang them about the doorways, or place them in the dog kennel and where flies gather. Other means include fly ribbon strips and bug electrocutors.

Q. What are some ways to cut down on mosquitoes without resorting to poisons?

A. Creating a favorable natural balance can do much toward eliminating the mosquito problem. Cleaning up breeding places, introducing predators and parasites—such as dragonflies, toads, frogs, snakes and certain aquatic life, and maintaining a healthful, unpoisoned environment for beneficial wildlife have all done more in the long run to fight this problem than sprays which kill everything indiscriminately, leave the conditions for more trouble unchanged, and eventually bring about resistance and immunity in the pests.

Sassafras repels mosquitoes as do garden pools with goldfish in them. Goldfish eat mosquito larvae. The castor bean plant located near porches, or marshy spots will reduce breeding.

Look for standing water and eliminate it if practical. Here are USDA recommendations for eliminating possible breeding places. Remove unneeded, temporary water containers. Flatten or dispose of tin cans. Place discarded bicycle and automobile tires in places where water cannot get into them. Fill in tree holes with concrete. See that cisterns, cesspools, septic tanks, fire barrels, rain barrels, and tubs in which water is stored are tightly covered. Empty and thoroughly wash birdbaths and pans for watering chicks at least once a week. Clean out rain gutters. Examine flat roofs after rains; be sure no water remains on them. Drain or fill in stagnant pools and swampy places. If pools cannot be drained or filled in, remove debris and floating vegetation. Check stock-watering tanks for larvae or pupae once each week or oftener; keep them clean; repair leaks. Fill around watering devices to prevent water from standing. Gravel or sand may be used. Examine philodendron and other plants in water in the house. Plants potted in soil will not breed mosquitoes, but saucers under the plants may accumulate enough water for breeding. Examine fish bowls and aquariums for larvae.

Protecting Trees

V

Q. What are cankerworms?

A. Cankerworms are green, brown or black worms a little thicker than a pencil lead and about an inch long. They do extensive damage to tree foliage, particularly maple, oak and elm. When disturbed, the worm drops from the leaf on which it's feeding and swings at the end of a silk-like thread. When it reaches the ground, it may climb another tree, bush or building. Cankerworm eggs are laid in the bark by wingless females who crawl up the tree trunks in late fall.

Q. Is there a weapon against cankerworms?

A. Yes, a sticky band of "tanglefoot" material applied to the tree trunk before November 1 is an effective way to control the pest damage by preventing the egg-laying female's climb. Trik-o is also recommended as a control measure.

Q. What is tanglefoot and how does it help prevent insect damage to trees?

A. Tanglefoot is a sticky compound which, when painted in a band around a tree, will prevent insects from going up the trunk. It can be purchased at seed or hardware stores. If it is not available, fly tanglefoot can be purchased and put around the trees in the form of a girdle.

Q. Please explain how scale insects injure trees and how they may be controlled?

A. Scale insects are tiny sap suckers that can occur in numbers large enough to weaken or kill an entire ornamental, shade or fruit tree. The pests secrete a shell-like armor mound around their bodies and around their feeding areas, a natural defense which makes scale insects difficult to control. The protective armor looks like encrustations on twigs and branches.

A highly refined "superior" or "supreme" type of dormant emulsifiable oil, which is available commercially, should be sprayed on trees before the buds break open. Cover the entire tree with the spray; avoid application if freezing temperatures are likely to follow. Natural predators—especially ladybugs—are also valuable in control of scale because they feed heavily on both scale eggs, larvae and adults.

Q. Is there an alternative to using poisonous sprays to reduce the concentration and damage of tent caterpillars in orchards and shade trees?

A. You can lessen the population of tent caterpillars by acting early in the spring. Remove the brown egg masses from the branches of any wild cherry trees in vacant lots and other areas. The moths lay these eggs during the summer on wild cherry and related trees. The cylindrical eggs are glued together in masses of 200 to 300 and form a characteristic brown band around twigs of the host tree. Every time a single egg mass is destroyed the potential threat of 200 to 300 more tent caterpillars the next spring is gone.

Q. Is there any way to control the bagworms that infect evergreens?

A. Yes, bagworms favor arborvitae, junipers and similar ornamentals, and are removed by hand fairly easily. By September they have stopped feeding and are pupating in their cone-like bags. Later, adults will emerge and the wingless females will lay eggs in these bags. To control bagworms at this time, pick the bags off the bushes or trees and burn them.

Trik-o is another effective means of controlling these pests.

Q. Last year I harvested my first crop of MCINTOSH apples from my young dwarf tree (40 perfect specimens). This year, I decided to go organic and just applied a dormant spray.

Now a problem appeared early in the summer. I noticed brown holes in the apples which contained small worms. My agricultural extension agent identified them as apple maggots. He told me that it was next to impossible to grow apples without a spray program unless I was the only one in my area growing apples. I don't demand perfect fruit, and a few blemishes wouldn't worry me, but the maggots make the apples unappetizing. I do not wish to go back to spraying, so if you have any suggestions as to what I could do next year, I would be very grateful.

A. You don't have to wait until next year to do something about the apple maggot problem. In fact, the sooner you start, the better. There are a number of things you can do, starting with the clean cultivation of your land—that is, collecting dropped apples and windfalls. Dropped fruits of the summer varieties should be collected twice a week; those of later varieties, once a week. These apples will be fed to livestock, dumped into water, or, if the quality is high, used for cider. If you have enough land to support them, a chicken or two and maybe a duck (as long as there are only a few trees involved) will help with the clean-up operation. The chickens will go after the windfalls and the worms, and the ducks will take after weeds and keep a great many apple maggot pupae from overwintering in the soil.

Another solution for the small plot gardener is bait traps similar to those used for codling moths. There are two formulas: one part blackstrap molasses, nine parts water and one cake of yeast per gallon of mixture; or substitute one part diamalt for the blackstrap molasses in the mixture. For 10 traps, add one quart of molasses or diamalt to nine quarts of water. Dissolve the yeast in water, add to

163

the mixture and stir well. Set the mixture aside for 48 hours, or until fermentation is complete. Fill wide-mouth quart jars with a plate of $\frac{1}{4}$-inch mesh hardware cloth in place of the regular canning plate. Secure a pulley, ring or screw eye to a branch of the tree and run a stout cord through it in such a way that it will clear the branches when raising or lowering.

Yet another approach, if the infestation is quite serious, is spraying with rotenone—a plant-derived material—from May through August. The non-chemical spray kills other insects and certain external parasites of animals as well as apple maggot. The non-chemical extract, however, has very little residual effect and its period of protection is short; hence the May through August applications.

Pesticides

VI

Q. Can insects really be controlled without insecticides?
A. Many successful organic gardeners have found that, with a little know-how, they are able to control most of the troublesome bugs—without investing a lot of time and money and without using poisons. By concentrating on improving their soils, organic gardeners have demonstrated that they can produce vigorous plants with real resistance to pests and disease.

Although organic methods don't provide 100 per cent control, organic gardeners are content to let the few pests they may find in their gardens have a small share. At the same time, there are important insect control methods that should be learned and used. Knowing just what a certain bug likes to eat and where the adult female lays her eggs is information that can be used effectively in ridding your garden of that enemy.

Q. How does an insect control program work?

A. With a little practice, you can learn to recognize at a glance the signs and symptoms of common pests. Once you've found out what the destroyer is, concentrate on methods to eradicate it.

The first and simplest thing to do is to try using seed varieties which are most resistant to disease in your region. Before investing in a large stock of any special seed, however, make certain that the variety will grow well in your type of soil. Your local county agent or state experimental station can advise you on this.

The second technique of insect and disease control is quarantine. It is dangerous to keep a diseased plant in a garden. People are apt to handle the leaves of the sick plant and then go on to touch healthy plants, thereby infecting the entire garden. There is a strong case for isolation and destruction of a pest after it gets established in an area. This calls for traps, parasites, safe inoculants and other natural methods.

Q. What can be used to remove insecticides and other poisonous substances from food?

A. Insecticides today are applied in bases which are not soluble in water. Many insecticides are systemic —that is they are injected into the plant or put into the soil to be taken up by the plant. In addition to insecticides, most fresh foods are treated with waxes, preservatives and dyes, none of which can be removed and all of which are suspected of being harmful. A solution of distilled vinegar and water is said to remove up to 85 per cent of surface residues from insecticides if the fruit or vegetable is left immersed in it for five minutes.

Miscellaneous

VII

Q. What is diatomaceous earth and how does it affect pests?

A. Diatomaceous earth consists of the skeletal remains of microscopic marine life deposited in lake bottoms millions of years ago. It is used as the basis for a non-poisonous insect-control formula developed a few years ago by a firm in Phoenix, Arizona. The substance releases microscopic spears of silica which cut into pests and allow body fluids to escape. Government and university tests have shown it especially effective against pests of stored grains and for control of some household insects.

Q. What is a trichogramma?

A. Trichogramma is a big word for a very small insect (you need a magnifying glass to see it). Yet this little fellow is taking a place along with the ladybug, praying mantis and other parasites or predators as one of our commercially important beneficial insects. The female trichogramma lays her eggs inside the eggs of harmful insects by use of a pointed egg-laying apparatus. When the trichogramma egg hatches, the young parasite eats the contents of the egg in which it lives, preventing it from producing a harmful insect. Primarily, the trichogramma attacks worms of the Lepidoptera order. Some of these are the codling moth, cotton bollworm (corn worm), European corn borer, looper, Oriental fruit moth, pecan nut case bearer, greenhouse leaf tier, imported cabbage worm and tomato worm.

Q. How do most insects manage to live through the winter in cold or freezing sections of the country?

A. Since few bugs are able to migrate to warmer climates, most of these cold-blooded creatures must

166

adjust to freezing temperatures—or die. Specially endowed insects like the chinch bug produce an antifreeze chemical that keeps their insides from turning to ice. Others calmly freeze without injury and await the spring thaw. On the other hand, the Cecropia silkworm moth spins a Thermopane-type cocoon that traps air between double walls for maximum insulation. Some insects dig down below the frost line and sleep away the winter.

Insects sense the approach of winter with a built-in "clock" geared to the seasonal variations in darkness and light. Known as photoperiodism, the phenomenon serves insects as a vital early-warning system. Long before winter, for example, the female grasshopper buries a mass of eggs wrapped in a gluelike jacket. Warmth will not hatch the eggs unless they have first been frozen—nature's way of making sure a late warm spell in autumn won't bring out baby grasshoppers to starve. In autumn, winged ants and ladybird beetles in California fly up into the mountains to spend the winter huddled by tens of thousands in wingless aeries. Many mosquitoes pass the winter as larvae, content to be frozen in ponds. When spring comes they thaw out and after metamorphosis, buzz off. Probably the most coddled of all wintering bugs is the corn-root aphid. Its eggs are carefully collected by a species of ant and carried to nests below the frost line. In spring the eggs are taken to the roots of early weeds to hatch.

CHAPTER 12

ANIMAL PESTS AND THEIR CONTROL

Even the most dedicated animal lovers can feel their blood boil when they look out on a vegetable garden ravaged by groundhogs or a lawn that has been corrugated by moles. Since all animals are part of the balance of nature, a status quo which organic gardeners are anxious to maintain, the trick is to discourage animals from attacking domestic plantings without harming the animals. And it can be done.

Over the years, observant gardeners have shown amazing resourcefulness in using everyday household items that are perfectly harmless to pesty animals, yet repugnant to them. You need go no farther than your kitchen cupboard to find what it takes to keep dogs away from your shrubs, cats out of your flowerbeds and moles from under your lawn. When animals are a problem, think deterrent, not poison.

Q. Is there any way, short of killing, to discourage deer from invading a garden?

A. Dried blood is one of the most effective deer repellents. It can be shaken right on foliage without danger of burning the plant. When it rains the dried blood washes down into the ground and feeds the roots. There is no odor from this method, provided the blood stays dry in storage; otherwise the stench will be acute. This inexpensive treatment should be applied about four times every spring.

Spent mushroom compost is also a popular deer repellent.

Q. How can groundhogs be kept from demolishing a vegetable garden?

A. They are almost sure to be discouraged from visiting any garden by the simple application of spent mushroom soil.

Q. What can be used to keep rabbits from chewing-up gardens and trees?

A. Rabbits are strictly vegetarians. They will not touch anything that has meat or blood on it. Put a spoonful of dried blood in about two gallons of water and spray garden plants and tree trunks with the solution. Rabbits will soon go elsewhere.

Q. When the neighbors' cats and dogs select your flower beds and shrubs to visit, is there a harmless way to get rid of them?

A. If cats and dogs visit your shrubs or flower beds, scatter some hot pepper in the area. They don't like it! (Cayenne pepper, garlic and other "strong" herbs or vegetables also make great non-poisonous bug-chasing sprays.)

Q. How much harm do moles actually cause when they invade a lawn? Do they eat grass roots and bulbs?

A. When obnoxious ridges appear just under the surface of your lawn, you're probably getting a visit from a family of moles. Although the ridges in themselves are annoying and unsightly, the problem is compounded by mice who consider the mole tunnels as an invitation to take up residence with you.

Contrary to popular belief, moles do not feed on grass roots or bulbs and other vegetable matter but are ever in search of grubs, worms, and other soil insects. The partially eaten bulbs found in mole runs are left there by other rodents such as field mice who use the runs for their convenience.

Q. Will planting certain flowers discourage gophers and moles?

A. Planting a border of scillas sends them packing and also gives the added benefit of lovely blooms in the spring. Castor beans are also effective in discouraging gophers.

Q. What are the most effective, most humane ways to get rid of moles?

A. Many home owners have achieved successful control through the use of commercial traps set into the tunnels. Naturally these must be checked and emptied daily.

One *Organic Gardening* reader heard that the castor bean plant repels moles, but didn't want to expose her young children to the poisonous beans. So she prepared an emulsion of castor oil and liquid detergent in a blender, two ounces oil to one of detergent. When it became as thick as shaving cream, she added an equal volume of water to the mixture and whipped it again. She put about two tablespoonfuls of the oil mixture into a regular sprinkling can filled with warm water, stirred it, and applied to the areas of heaviest mole infestation.

Q. How can damage done to trees by red squirrels be recognized?

A. The winter diet of the red squirrel includes the buds of several kinds of conifers. Although their work is not very noticeable outwardly, the effect on tree growth and form can be serious. The big buds of Scotch pine are cut off, the green contents eaten and the bud coverings dropped on the ground or snow. Occasional damage to white and red pine is similar, with a small section of the branch or leader being nipped off along with the bud. Squirrels will cut off twig tips up to ten inches long from a Norway, white spruce, balsam or European larch. They carry them to a resting place where the buds can be eaten out. Sometimes the terminal bud of a spruce leader is consumed through a hole chewed in the base and left in place, a type of damage often mistaken for an attack by the white pine weevil.

Q. What can be done to stop red squirrel damage?

A. While certain commercial or homemade repellents help curb squirrel raids in the garden (interplanting onions is also said to be a good measure), they can't do much for trees. A type of white-mesh trunk guard provides some assist. Trapping is effective. A box trap baited with an apple works well. Traps should be set in a pen with the bait concealed from above to avoid catching birds. Try cage-type traps which capture animal pests without harming them. Some common hawks, such as the goshawk, feed on red squirrels.

Q. Do skunks damage the garden?

A. Detailed diet studies of skunks have shown that their principal food is insects, mostly types which are injurious to plant life. They are especially good at digging out the June bug or May beetle in both larval and adult stages. Mice constitute another important food item, as do fruits when ripe and scattered on the ground surface.

A skunk will take an occasional bird, usually one which is already dead or injured. In the fall, skunks sometimes cause damage to beehives in their search for food. Since they are mainly ground-feeders, this difficulty can be prevented by placing hives on benches or stands. Despite their common reputation, the species is definitely more helpful than harmful.

Q. How can rabbits and field mice be kept from damaging fruit trees?

A. Rabbits will chew the tender bark and sometimes the lower branches from young fruit trees—plum, peach, cherry and first-choice apple. When trees are completely girdled, they seldom survive. Fruit trees should be protected from rabbits every winter from the time they are planted until they are old enough to bear fruit. One method is to put a cylinder of one-fourth inch hardware mesh around the trunks from slightly beneath the soil line to the lower branches. You won't have to bother with the mesh again for four or five years if you leave it large enough for the trees to grow.

Another successful method is to wrap strips of aluminum foil around the trunks. This also protects them from sunscald during the warm days of winter. Lower branches can be wrapped in the same manner, and the foil expands as the tree grows.

Meadow mice nesting in mulch or grass at the base of the trees will also feed on the bark. To reduce the possibility of damage, pull any grass or mulch up to at least two feet away from the trunks.

Q. What can be done about the nightly raccoon raids that many gardeners have to contend with?

A. Raccoons like to work in the dark so a floodlight in the garden is one way of deterring them. Another is a fence of chicken wire, stapled to posts that allow the wire to project a foot or so above them. Since raccoons climb fences rather than dig under them, they will find the fence difficult to get over because their weight pulls the slack top over them.

CHAPTER 13

THE BIRDS AND THE BEES

Both birds and bees rank high among the important creatures who assist the gardener in his quest for perfection.

A wise gardener does whatever he can to attract birds, for he knows they help to control destructive insects in the garden. Explore the many ways organic gardeners have devised to attract birds and keep them around for as much of the year as possible. Certain plants are particularly attractive to birds, and these species should be kept in mind when planning your garden.

Bees are a must for any orchardist. He will take great care to encourage their presence by avoiding the use of chemical insecticides, all of which are toxic to bees. Many fruit growers keep bee hives, not for the honey crop, but as cross-pollination insurance.

Always be aware of the priceless services the birds and bees render, and make your garden a place they will be reluctant to leave.

Birds

I

Q. Why are birds important to a gardener?

A. One of the most successful and cheapest ways to control insects in the garden and on farms is to get a variety of birds to do much of the work. Sixty or more helpful species of birds can be attracted in

Place bird feeders in a strategic part of the garden.

virtually any agricultural area of the United States.
Q. What are the best ways to attract birds?
A. Put out feed for them all year. In spring and sum-

mer, provide suitable nesting areas and keep some nesting materials nearby. While birds are very resourceful, a few handy strings, rags and feathers will be gratefully used. Provide water for the birds, especially near nesting times, so the parents will not need to leave the eggs or the baby birds to search for water.

Q. What foods can I provide that will encourage the birds to come early and stay late?

A. Bread crumbs, beef suet, orange and apple slices, dried currants, raisins, cracked corn, sunflower seeds, and a mixture of millet and hemp seed are all effective enticements. Good use can also be made of dried melon and pumpkin seeds, dried baked goods, rolled oats and crushed egg shells.

Q. Do some plants attract birds more than others?

A. Yes, sunflowers, cosmos, marigolds, asters or California poppies will encourage many weed-seed consumers to remain around a garden. A hedge of multiflora or Rosa Rugosa roses, Japanese barberry

Woodpecker eating suet from lobster bait bag. Feeder made from birch log is suspended from the wire.

hedge, bush cherry, bitter sweet, Michigan holly, highbush cranberry, bush honeysuckle, snowberry, and many other kinds of shrubs offer equally good food for birds.

Keep the birds in mind when placing trees or specimen plantings. Many attractive landscaping trees will supply nesting sites for birds and often food as well. Sugar maple, flowering crab, the nut trees and other fruit trees serve a double purpose. Junipers, mountain ash, and the hollies are a few others especially recommended for attracting birds.

Q. Do woodpeckers injure trees?

A. These birds might be regarded as a mixed blessing, although much more can be said for their virtues than unintentional vices. A woodpecker's drumlike pecking may poke holes in the bark through which tree disease organisms could enter. However he's really after some very destructive insects, like weevils, May beetles or wood borers—and seeks out trees infested with them.

Q. What flowers will attract hummingbirds to a garden?

A. Among flowering favorites of hummingbirds are honeysuckle, phlox, day lilies, nocotiana, cardinal flowers and trumpet vine. They're also fond of the mint herb known as bee balm, which attracts pollinating bees as well.

Bees

II

Q. How do insecticides affect bees and the honey they produce?

A. All commonly used insecticides are toxic to bees. Those that easily drift or are applied from spray planes cause the most harm, frequently wiping out colonies across a wide area. Particularly lethal are any of the arsenical insecticides often applied to crops. U. S. Department of Agriculture analyses disclose that as little as one-third of a part per million of the bee's body weight is enough to cause death. Their tests also show that any arsenical carried into the hive with pollen on the bee's legs and stored for future food remains poisonous for months.

One would expect that some of the enormous quantities of insecticides applied on cultivated crops would show up in honey. That is not the case. Nectar is carried in the honey sac, a specialized part of the alimentary tract. When the nectar contains poison, the carrier is quickly affected. Instead of returning to the hive, the bee attempts to throw off the effect of the poison and becomes lost or dies in the field.

If a bee does manage to return with a load of poisoned nectar, there is a second safety factor. Every drop of the nectar is rehandled by the hive bees, which are exposed to poison longer than the field bees. Hive bees tend to leave the colony when poisoned, carrying with them the toxic nectar. It is therefore unlikely that such poisons would ever be stored with honey.

Q. What plants do bees gather the best honey-making nectar from?

A. Nearly three-fourths of the commercial honey crop comes from alfalfa, clover, buckwheat, cotton and orange blossoms. Although bees depend on apples and other orchard fruit trees, plus dandelion, mustard and goldenrod for most of the nectar they need to maintain their hives, they store very little of this in the comb cells.

Q. Are bees prone to any insect diseases?

A. Honeybees are normally quite healthy, but they are subject to a disease which accounts for nearly all of the losses in honey and beeswax production. It is a bacterial plague called American foul brood. In many areas, authorities examine hives in order to destroy infected bees and prevent the spread of this highly contagious disease. Sound beekeeping practices, especially strict sanitation, clean equipment and quick isolation of infected bees are the most effective control measures.

Q. Will freezing honey prevent darkening and crystallization?

A. Yes. Honey kept on the kitchen shelf gets sugary and turns dark after a while. Putting it in the refrigerator doesn't help. In fact, crystallization occurs sooner. None of this happens to honey kept in a freezer. Store it there and remove a one to two week supply at a time.

Q. Are there any plants which are particularly attractive to bees and can be used to entice them to a garden?

A. Yes, bergamot (bee balm), and lemon balm attract bees for faster pollination. Thyme and borage are also very effective bee baits.

CHAPTER 14

PLANTS

Some people seem to have a green thumb when it comes to plants. Their roses bloom beautifully, untroubled by either mildew or black spot. The leaves on their house plants stay a strong healthy green. And it seems that the boldest mealy bug (one of the most common enemies of the indoor plants) wouldn't dare to attack their aspidistra.

Actually, what appears to be the proverbial green thumb generally turns out to be good, old-fashioned know-how. These people are still doing things as they were done before the chemical age took over. They know that the loveliest roses grow without pesticides and the healthiest house plants thrive on organic fertilizer. Just about any of the diseases that plants fall heir to can be cured or controlled by natural methods.

House Plants

I

Q. The new leaves on our house plants sometimes get smaller and turn light green. What can be done about it?

A. Feed the plants. If new leaves are progressively smaller and paler than older foliage, it's a sign that fertilizer is needed. Although most house plants are overfed—few require more than three or four feedings a year—the timing and applications are important. Potted plants should receive only as much fertilizer as they can incorporate into the food manufactured in their leaves. During winter when they get little light, less fertilizer is needed. In the long days of summer they use more. Young, actively-growing plants take more feeding than older, more mature ones. Good pot-plant fertilizers include fish emulsion, compost or manure "tea."

Q. Some ornamental plants seem to dry out too fast no matter how often they are watered. Is there any remedy for this trouble?

A. Moisture retention can be improved in several ways. In the potting mixture itself, incorporate some peat moss, vermiculite or other highly absorbent material to help hold water for use as the plants need it. Check to see that drainage is adequate, but not excessive. A clay pot can be kept from drying out rapidly by setting it inside a nonporous container—one large enough to leave an air space of about an inch all around the clay pot walls. This will not interfere with necessary evaporation and "breathing" but will better resist the drying effects of hot wind, sun or overheated houses.

Plants should not be over-potted. If the pot is too large, the plant may fail because it cannot absorb the water in the soil quickly enough. This danger is minimized in a clay pot with its porous walls. But there is more aeration—and therefore more soil cooling—if the plant is in a pot that comfortably fits its roots so that it uses the moisture content at a better rate.

Q. What potting soil mixture should be used for house ferns?

A. A mixture of four parts loam, four parts leaf mold, two parts sand, and two parts dried manure is recommended for ferns. A half-cup of bone meal and two cups of broken charcoal (one-fourth inch or less) added to each peck of the mixture will improve it. This soil is also very suitable for African violets and begonias.

House Plants.

Q. Is there any validity to the theory that house plants will grow better if an eggshell is left on top of the soil overnight in a flower pot?

A. The theory is partly right. In addition to their principal ingredient, calcium, eggshells contain over one per cent nitrogen and about 0.4 per cent phosphoric acid. These are all, of course, important nutrients for any plants—including house plants. However, much better than the overnight shell-placing method would be crushing or grinding the eggshells and adding this to the plants' soil or fertilizing mixture.

Q. Which is best for light-grown indoor plants: fluorescent bulbs, sun lamps, or incandescent bulbs?

A. Incandescent bulbs give off red rays that are beneficial to plants. Although they give off more heat than blue-ray fluorescents, some growers add two

25-watt incandescents to each pair of 40-watt fluorescents. Some house-plant specialists advise using incandescents as an adjunct to regular indoor gardening, pointing out that the added illumination often helps overcome spindliness from lack of natural light. (Fluorescents developed especially for house plants, such as the Sylvania Gro-Lux lamp, combine the blue and red light rays effectively.) A sun lamp should never be used on plants. Its rays are detrimental to them.

Q. Is softened water suitable for indoor and greenhouse plants?

A. Water softeners operating on an ion-exchange principle produced water injurious to plants by causing a sodium build-up in the soil. Use untreated tap water for all plants or—better still—rainwater, spring or well water whenever available.

Q. What indoor plants can do well where the rooms become very warm and don't receive much light?

A. Favored plants for locations that get lots of heat and little light include the Aspidistra elatior or "cast-iron" plant, the Aglaonema or "Chinese evergreen," and the Sansevieria or "snake plant." All these are tolerant of overheated homes and high summer temperatures. Guard against overwatering; keep them preferably in clay pots.

Q. What can be done to rid indoor plants of the mealy bug?

A. This is one of the indoor plant's most annoying and harmful pests. Oval, with short projections from its body, the mealy bug often looks like bits of cotton fluff because the female carries her eggs in a cottony sac. Mealy bugs are sluggish, move very little. Dip a cotton-tipped toothpick in denatured alcohol and lift off each visible bug, repeating once a week until no more can be seen. A swab of cotton soaked with alcohol can also be used to clean off

Potted citrus fruits do well indoors.

leaves or stems of plants. Miscible oil spray solutions
are also reported effective.

Q. What is the rule for fertilizing foliage plants?

A. Little fertilization of established foliage plantings
is necessary. With many foliage plants you're interested only in maintaining the plants, rather than

getting any large amount of growth. In fact, once plants are well established, new growth may not be desirable at all if it is soft, spindly, rank or hard to manage.

An application of fertilizer twice a year may be quite satisfactory for foliage plants, particularly if only one or two new leaves a year are desirable.

Q. Can you name any fruits and vegetables particularly suited to growing indoors?

A. Try tomatoes and peppers. Both popular salad ingredients are classed as annual vegetables, though they're really fruits—and in their native tropical habitat both of them are perennials. Ordinary bell peppers grow well as house plants in eight-to ten-inch pots, staying healthy and bearing flowers and fruit around the calendar. Dwarf Italian sweet peppers, or the hot Mexican types, are also sold as ornamental or Christmas peppers, usually perform well inside. Start some seed or dig up a stocky plant or two this autumn to pot up and try the peppers-on-the-sill plan. Do the same thing with a stout late tomato from the September garden, or give a midget cherry variety such as *Tiny Tim* a chance to produce indoors—perhaps in a decorative hanging basket.

Most popular and widely planted among the edible ornamentals are indoor versions of several citrus trees. Their glossy leaves, fragrant blossoms, and ability to ripen fruit during any season make them deserving favorites. The *Ponderosa* lemon is probably the best-known window-grown house plant. Close behind are the *Chinotto, King* and *Sweet* orange. A newcomer is the *Otaheite*, an attractive white-and-pink-blossomed dwarf orange that sets prettily in a five-inch pot, bears up to a dozen ¼-normal-sized fruits at a time. The dwarf tangerine (*Citrus nobilis deliciosa*) not only has handsome foliage and delightfully aromatic flowers, but the bright-colored fruit is fine eating.

Other citruses suited for house-bound culture include the *Meyer* lemon, limes, kumquats, grapefruit and calamondins. All are attractive and doubly useful.

Q. What are the special needs of indoor edible ornamentals?

A. The biggest factor in getting good results with any of the edible ornamentals is good care. It's important to realize that nearly all are tropicals, at least by birth, and that the closer we can come to providing their natural environment, the better we'll do with them. First, remember that potted plants can't send their roots very far after food; it must be brought to them. Do this by potting in compost-enriched soil; make sure it is loose, porous and well-balanced in plant nutrients. Just as with the outdoor garden, a sound idea is to test it, add some finely ground potash and phosphate rock or a little bone meal when mineral deficiencies are indicated.

In soil to be used for repotting, incorporate generous amounts of compost. If you're not repotting, but find a boost is needed, wash out the top inch of soil and replace it with a humus-rich material. Manure or compost-water tea and diluted fish emulsion are helpful liquid fertilizers for a periodic stimulus to plant growth.

Next to soil, the most important thing is exposure. Fruiting plants need all the sun they can get. Some fruit can be expected in a partially sunny location, but heaviest bearing will result from full sun. Temperatures, too, are a vital influence. For an ideal growing climate, the daytime reading should range from about 65 to 75 degrees, and should slide down to 55 to 60 overnight. The night drop is important because it allows the plants to rest and mature the growth made during the day.

When it comes to water, too much can be worse

than not enough. More house plants are drowned by well-intended overwatering than are ever harmed by drought. Of course, no plant should ever be allowed to suffer from lack of moisture, particularly those tropicals that you want to coax into flowering and fruiting. Keep the air as humid as possible, and water liberally as soon as the surface soil starts to dry out. This is much better than giving small amounts of water more frequently. Also, be sure to water-spray the foliage and all top growth at regular intervals to increase humidity, keep the plants clean, and discourage insect pests.

Roses

II

Q. What is black spot?

A. U.S.D.A. booklet, "Roses for the Home," carried this description of black spot: The circular black spots have irregular radiating margins and are frequently surrounded by a yellow halo. The spots may be as small as 1/16 inch in diameter or they may nearly equal the width of the leaves. Infected leaves characteristically turn yellow and fall prematurely. When the attack is severe, the plants may be almost completely defoliated by midsummer.

Q. What causes black spot to infect roses?

A. Black spot is caused by a fungus which grows into the leaf and forms the familiar black spots by which the disease is known. A few days after the black spots are formed little black pimples show up in the spots; these indicate the spores are about to be discharged. Therefore, any infected leaf should immediately be picked and burned or composted. Otherwise the spores may be carried by rain and wind on the gardener's tools, hands or clothing, or even by insects, and thus healthy leaves may be quickly infected.

Q. How is black spot spread?

A. Black spot is spread by splashing water and infection takes place only when water remains on the leaves for several hours. Consequently, black spot is most serious in areas of high rainfall and is least serious in arid regions. Even in dry regions overhead irrigation may permit the development and spread of this disease if viable spores are present.

Q. What suggestions can you give for defending against black spot?

A. The best defense is a strong offense. Practice prevention by pruning all diseased leaves off plants in early spring and raking away foliage that may serve to overwinter spores. Full sun, thick mulch and adequate ventilation is the best preventive for black spot. Remember not to water rose bushes toward evening. Check catalogs for varieties with most disease resistance.

Q. What is powdery mildew?

A. A fungus growth that covers tops of rose leaves and young shoots with white, powdery spores, mildew ranks second among common rose troubles, following only the nefarious black spot. When attacked, leaves often become twisted, later may turn red or yellow. Severe infection kills the plant's growing tip, keeps buds from opening and causes leaves to drop prematurely.

Powdery mildew is easily diagnosed. The fungus (Spaerotheca pannosa var: rosae) affects only roses, while mildew diseases of other plants do not spread to roses. The fuzzy, cobweb-like growth occurs frequently during damp seasons, although it is not spread by splashing water (as is black spot) and the spores actually do not germinate readily when wet. It also differs from most other disease orga-

Straw makes a good mulch for rose bushes.

nisms in that it grows on the surface rather than inside the host plant. Temperatures between 64 and 75 degrees favor rapid development of the tiny wind-carried spores. Succulent new plant leaves or shoots can be covered with mildew in a short time.

Q. Are there ways of preventing powdery mildew?

A. Plenty of rich natural fertilizer boosts plant health and resistance. Be sure to water thoroughly and at the right time (never in the evening)—generally once a week if there's been no rain.

Added points to keep in mind: Fall sanitation is vital. Clean up all leaves that may harbor spores which overwinter and may start the trouble cycle again in spring. Look for hardier, less susceptible roses. Hybrid teas, climbers and ramblers "catch" powdery mildew far more often and quickly than Wichuraina, Welch multiflora and Rugosa types.

Q. How can it be controlled when it does come?

A. Control measures for powdery mildew start with prompt removal of all infected parts. Collect and destroy leaves and stems showing mildew as soon as it is noticed. The sooner the cleanup, the less chance spores have to travel and spread. Thin out the plants to let in sunshine and air; try to keep plants dry. Full sun and adequate ventilation comprise the best mildew preventive. Exposure to conditions such as low relative humidity and either very high or low temperatures will also kill spores.

Q. In planning a garden's landscape pattern is it advisable to include both flowering hybrid roses and hedge or fencing types? Are there any precautions necessary in using these together?

A. Bush-type roses, such as rugosas and multifloras, have strong growth characteristics and tend to overrun or crowd out the more delicate hybrid varieties if planted too close. It's better to allow these classes an attractive location by themselves.

Q. Will mulching keep winter freezing from damaging roses?

A. Most winter injury to roses comes from alternate freezing and thawing of soil during warm spells in winter. This problem can be minimized by proper application of a good mulch around plants to prevent the soil from freezing too deep and to act as an insulator.

Don't apply it too early, though, since doing that may prevent soil from freezing at all, thus allowing plants to sprout too soon in the spring—before danger of killing frosts has passed. Wait until the ground has partially frozen, then spread six to eight inches of material. If you already maintain a summer mulch, simply add to it for winter.

The best mulches are kinds that do not tend to pack down tightly. Coarse ground corncobs make an excellent mulch; others that work well include clean straw, wood chips, peat moss.

Annual/Biennial/Perennial

III

Q. What do gardeners mean when they classify flowering plants as annual, biennial and perennial?

A. An *annual* is any plant whose seed when sown in the spring will produce summer or fall blossoms and not live over the winter; (asters, begonias, carnations, larkspur, marigolds, morning glories, nasturtiums, petunias, poppies, snapdragons, sunflowers, sweet peas, zinnias, etc.);

a *biennial* is a plant which normally requires two years to complete its life cycle (blooming the second season after seed is sown) before it produces seed and dies. (Canterbury Bells, English daisies, forget-me-nots, foxglove, hollyhocks, pansies, Rose Campion, Sweet Williams, etc.);

a *perennial* is any plant which lives for more than two years; (bleeding heart, border chrysanthemums, day-lilies, dictamnus, goldenrod, Japanese anemone, paniculate phlox, peony, plantam lilies, etc.)

Q. Are plants grown from bulbs considered annuals or perennials?

A. Most annuals are grown directly from seed, but some plants are grown from tubers or bulbs. For example, crocus and lilies are actually perennials that die back each year to the ground while the roots remain alive. This mixture earns them the name "false annuals."

Q. Do annuals have any advantages over other flowering plants?

A. Yes. Most are easy to handle, inexpensive, ideal for temporary plantings, fine as fill-ins after perennials have stopped blooming, and offer a color range from pure white to almost black.

As cut flowers, annuals are almost indispensable. They are used for bedding plants, edgings, for rock garden subjects, as climbers for covering trellises and arbors. Included in the annual vines class are the morning glory and ornamental gourds, the common hop, thunbergia, hyacinth bean and balloon vine.

Q. For best results, how should different types of annuals be handled?

A. *Hardy*—may be sown outdoors before frosts have entirely ceased right where they are to grow. Generally sown from February to May, some hardy annuals such as sweet peas can be sown in autumn. Often the early seeding is done in a semi-protected area, such as along a fence or wall, and the seedlings are later transplanted to the desired location.

Half-hardy—usually sown before full warm weather; often these are planted inside in February or March, as half-hardy annuals need warmth to get a good start. Once established, they are quite hardy in the garden.

Tender—seed started in the house or greenhouse, as plants require more warmth than the half-hardy group; a temperature range of 60° to 70° is considered correct; be sure that seedlings get enough light and are not overcrowded; common practice is to transplant seedlings into small pots in which they later can be placed directly in ground.

Most annuals do best in an open, sunny location, that has added large amounts of manure, leaf mold, etc. worked into the soil.

Q. When is the best time to plant biennials?

A. There are two groups of biennials. One group, known as the true biennials, includes Canterbury Bells, Sweet Williams, foxglove, hollyhocks, and Rose Campion. These are best sown in mid-June to mid-July to obtain healthy vigorous plants the following year.

The second group of biennials is comprised of pansies, forget-me-nots, English daisies and English Wallflower. Generally it is best to sow this group in August to avoid danger of winter-killing. Since biennials in this group are not very hardy, it's a good practice to protect them with a coldframe in regions where winter temperatures drop below 20°F. In any case, all biennials should be protected with a heavy mulch after the ground has frozen from weather extremes and heaving during warm spells.

Q. Are there any special rules or cautions in planting biennials?

A. When starting biennials, prepare the soil as for any good seedbed, making the soil fine, moistening it, adding leaf mold, peat moss or other humus material. A good practice is to mix the tiny seeds with dry sand in order to get even distribution in the rows.

Be careful when watering the seeds; if they get too little water, they'll dry out; if too much, damping off may result.

Once the seedlings have about four to six leaves and can be easily handled, it's time to transplant. Water sufficiently until they are well established. Mulch to conserve moisture.

Q. What is the best soil for perennials?

A. Perennials are adapted to a wide variety of conditions. Some are found growing wild in wet spots; others thrive on rocky hillsides or dusty, gravelly soils; some in rich bottom-lands.

A rich soil, not too heavy, is best for practically all perennials. It should be light and crumbly, well-drained but moisture-retaining, amply supplied with humus.

The best way to provide this is to dig at least 18 inches deep and work in a three-inch layer of leaf mold or peat moss, plus a one-inch layer of rotted manure or compost; increase these amounts if your soil is poor to start with.

Q. What is the best way to start perennials?

A. The perennials that are true to species are most easily propagated by seed. They may be sown in spring or midsummer, in a well-prepared seedbed or in pots or flats in a cold frame or greenhouse. These will give strong plants by fall, plants that will bloom the following year. Aquilegias and other strong-germinating seeds may be sown where the plants are to grow. Seeds of a few perennials, such as phlox, must be sown immediately after ripening or they will not germinate. It's a good idea to have a special propagating bed somewhere in your garden for starting new plants and storing surplus ones.

Many durable perennials may be propagated by division of their mature clumps, and some from late summer cuttings from main shoots. Most perennials should be divided and reset every three or four years.

Q. When should they be transplanted?

A. Thin your seedlings as soon as they have their first true leaves, and give them plenty of room at all times. Transplanting is best done in the fall, at least four to six weeks before freezing weather is expected. Eight or ten weeks before is even better in order to let the plants get well established. Always water throughly when transplanting.

Q. Do perennials need winter protection?

A. Mulching in late fall is recommended, even though the plants are hardy. After frost has penetrated the ground an inch or so, apply a mulch three to six inches thick, depending on the severity of your winters. It should be a light mulch, such as strawy manure, pine needles, fresh or partly decayed leaves, peat moss or salt hay. Be careful not to smother any plants that have evergreen leaves. Mulch under these, and use evergreen boughs to protect their upper portions.

Miscellaneous

IV

Q. What sort of plant-protecting covers can be used to avoid cold or wind damage in early spring?

A. One of the most common protectors is a specially treated paper-like cone resembling a hat. Various types of hampers, bushel baskets and wax-like paper containers are good improvised protectors. Recently various kinds of polyethylene covers have appeared on the market.

If paper caps are used, punch a small hole in the top to let moist air escape, especially after the weather becomes warmer. When day temperatures are above 55-60 degrees, caps can be removed. Keep them nearby so you can cover plants at night until late-frost danger has passed.

Q. How much sooner can outdoor plants be started if plant-protectors are used?

A. When plant protectors are used, you can usually plant a week or two earlier than otherwise possible. Added protection from late frosts and wind in early spring enables plants to become established more quickly and thus hastens maturity.

Protectors can be used on crops seeded directly in the garden, such as cucumbers, melons and squash. These crops can be sown a week to ten days earlier than usual and then covered with paper caps or cones. Transplanted crops such as tomatoes, peppers and eggplants also can be protected with covers, but require larger ones.

Q. What is meant by "forcing" a plant?

A. Forcing refers to the growing and blooming of plants in advance of their usual season. Flowering branches that are most commonly brought indoors for forcing include: forsythia, pussy willow, winter hazel, hardy jasmine, magnolias, flowering crabapple and dogwood.

The gardener who wants a continuous supply of blooms cuts a fresh lot every week or so from early February to late March. Use a sharp knife or shears and make a clean diagonal cut. To soften the hard outer covering of the buds and encourage movement of sap in the stems, the branches should be completely submerged in a tub of water for three or four hours. Then put them in deep containers of water and keep them in a cool dim place until the buds begin to show color. If the water in the containers is not changed daily use a piece of charcoal to keep it fresh.

Q. What is meant by "retarding" and how is this practice applied to plants?

A. Retarding is the opposite of forcing, and is commonly used to slow up the growth of plants like hydrangeas, Easter lilies and azaleas so they will flower on a later date. Cold storage may be used to retard Easter lilies while deep, unheated but covered pits or special houses where light and temperature can be controlled are common for retarding other bulbs and many plants. Darkness and a temperature around 40 degrees are essential. Sometimes a cool greenhouse where these factors can be managed is suitable for retarding. These methods can often extend the dormancy of plants for several weeks.

Q. How often should potted ornamentals be fed?

A. Infrequently. Plants grown in water-tight containers may be damaged by accumulation of soluble salts which occurs from too-frequent fertilizing, especially with concentrated or chemical fertilizer. Nutrients are best supplied by manure "tea," fish emulsion, compost or rock powders.

INDEX

Acid peats, 11
Acid soil, 13
Actinomycetes, 21
Aeration, soil, 11, 16-17, 73, 140
African violets, soil for, 180
Agriculture Department, U. S., 160, 177, 186
Alfalfa, 11, 13, 33, 177
Algae, 22
Alkaline soil, 10-11
All-purpose lawn, 121-22
Alsike, 13
Aluminum, 13, 31
American foul brood, 178
Anaerobic method of composting, 48-49
Angle worms, 75
Animal pests, control of, 168-72; cats, 169; deer, 168-69; dogs, 169; gophers, 169-70; groundhogs, 169; mice, 171-72; moles, 169-70; rabbits, 171-72; raccoons, 172; skunks, 171; squirrels, 170-71
Annuals, 190-93; advantages of, 190-91; definition of, 190; grown from bulbs, 190; half-hardy, 191, handling, 191, hardy, 191; tender, 191
Ants and anthills, 139-40, 144, 155
Aphids, 144, 155, 167
Apple maggots, 162-64
Aristotle, 73
Artificial chemical fertilizers, 28-29, 33, 41-42
Auger hole, making, 9
Austrian winter peas, 12

Bacteria, 1, 13, 21, 61
Bacterial diseases, 30-31
Bagworms, 144, 162
Balfour, Lady Eve, 3
Banana wastes, compost material, 56
Banks, grass used on, 128-29

Bark, shredded, as mulch, 107
Barn flies, 143
Bean beetles, 144-45, 154
Bean rust, overcoming crop loss to, 116-17
Bedstraw, 19
Bees, 173, 176-78; honey darkened and crystallized, prevention of, 178; insect diseases, 177-78; insecticides that affect, 176-77; plants providing best honeymaking nectar, 177; plants that attract, 178
Beetles, bean, 144-45, 154; cucumber, 147; Japanese, 149, 154; ladybird, 167; ladybugs, 144, 151; May, 171
Begonias, soil for, 180
Bents, 122, 127
Biennials, 190-93; definition of, 190; rules or cautions in planting, 192; time to plant, 191-92
Biological control of insects, 158-59
Birds, 173-76; attracting, 174-75; food to encourage, 175; hummingbirds, attracting to garden, 176; importance of, 173-74; plants that attract, 175-76; woodpeckers, injury to trees by, 176
Black spot, 186-87
Blood meal, 32
Blueberries, compost for, 55-56
Borers, corn, 147, 155; peach, 151
Boron, 31, 41
Bulbs, plants grown from, 190
Burr clover, 13

Cabbage butterflies, 156
Cabbage maggots, 145
Calcium, 10, 31
Canadian bluegrass, 127, 128
Canker worms, 94, 161
Carbohydrates, 35

Carrot roots, well-formed, method of getting, 116
Castor beans, 170
Caterpillars, tent, 145-46, 162
Cats, control of, 169
Cecropia silkworm moth, 167
Chemical fertilizers, 72, 74; artificial, 28-29, 33-34, 41-42; combining organic and, 42; hardpan formed by use of, 29-30; organic vs., 27-32; plant disease caused by, 30-31; soil aeration, interference with, 29
Chemical weed killers, 81
Chinch bugs, 140, 146, 154, 167
Chlorides, 30
Christen, Fred L., 8
Claypans, 7
Clay soils, 10
Clover, 8-10, 12, 13, 19, 33, 177
Coal ashes, compost material, 68-69
Cobalt, 41
Codling moths, 146-47
Coffee grounds, compost material, 59; as mulch, 104
Common Bermuda grass, 129
Compost (composting), 1-2, 3-4, 14, 32, 45-71; aerating pile, 63-64; aerobic method, 48-49; application of, 64-68; banana wastes in, 56; blueberry land materials, 55-56; coffee grounds in, 59; cotton materials in, 56-57; definition of, 45-47; do's and don'ts, 68-69; earthworm best for use in, 74-75; earthworm method, 49-50; elements necessary for, 47; 14-day method, 47-48; green matter in, 57-58; hair in, 57; heaps and containers, 59-64, 68-69, 75; importance of, 45-51; Indore method, 46, 47; lawn application of, 134; manure in, 54; materials, 52-59; municipal composting, advantages of, 70-71; pH range, 61; process on leaves, 51, 54; ring method of application, 67; seaweed and kelp in, 54-55; sheet method of, 49, 51; shredding, 52; Spanish moss in, 59; wood ashes in, 56
Compost activators, 68

Compost watering, 65
Containers, compost, 59-60; bins for, building, 61-63; garbage can, 63
Contour plowing, 25
Copper, 31, 41
Corn borers, 147, 155
Corn earworms, 147, 154, 155
Cotton, compost material, 11, 56-57
Cotton blossoms, 177
Cottonseed, 32
Cowpeas, 13
Crabgrass, elimination of, 81
Crimson clover, 13
Crop rotation, 15-16
Cucumber beetles, 147
Cultivation, seed, 83-84
Cutworms, 147-48, 155
Cyanogas, 140

Damping-off, 81; prevention of, 81-82
Dandelions, 19, 78
DDT, 72, 74, 150
Deer, control of, 168-69
Diatomaceous earth, 166
Diseased wastes, compost material, 69
Diseases, bacterial, 30-31; fungus, 30-31; lawn susceptibility to, conditions making, 137
Dogs, control of, 169
Drainage, 11, 137; proper, for planting trees, 96-97
Dusts (dusting), 2, 74, 143

Earwigs, 148, 155
Earp-Thomas, G. H., 3
Earthworm method of composting, 49-50
Earthworms, 1, 8, 42, 61, 72-76; castings of, 76; composting method, 49-50; compost pit, 74-75; population, encouraging large, 72, 74; role in soil fertility, 72-73; wintering of, 76
Elegant fine-bladed turf lawn, 122
Erosion, 11, 24-25
Evaporation, 37

Evergreens, fertilizing, 97, 98; seedlings of, planting, 97-98
Extra-sturdy working lawn, 122

Fennel, 111
Fermentation, compost, 61
Fertilizers (fertilization), 2, 27-44, 121; amounts for small gardens, 43; before mulching, 104; chemical, see Chemical fertilizers; evergreens, 97, 98; fish scraps, 35-36; fruit trees, 89-91; greensand, 43-44; ground rock, 27, 39-40; indoor plants, 183-84; manure, 36-38; natural, 28; nitrogen, 32-33, 34, 38; organic, see Organic fertilizers; phosphorus, 33, 34, 35, 38; potash, 33, 38; shade trees, 98; taboos, 41-43; trace elements, 40-41; types of, 27-28; vegetables, 118-20
Ferns, potting soil mixture used for, 180
Fescue, 129, 139
Field peas, 13
Fish meal, 32
Fish scraps, fertilizing value of, 35-36
Fire-blight on pear trees, combating, 94
Flies, 143, 153, 159
Flowers, compost application, 65; grouping by need for acidity in soil, 19
Fly ribbons, 143
Forest Service, U. S., 58
Fourteen-day method of composting, 47-48
Fruit trees, fertilizing, 89-91; reclaiming old, 93-94
Fungi, 1, 21-22, 95, 137, 139, 186-87; diseases, 30-31

Garbage can, compost container, 63
Garbage disposal, methods of, 70
Garden crops, manuring, 36
Gardening, organic, definition of, 1
Galium aparine, 19
Germination in seeds, obtaining highest percentage of, 84-85

Giant ragweed, 79
Glauconite, 43
Gophers, control of, 169-70
Grading of lawn, rate of slope in, 125
Grasshoppers, 148-49, 154, 167
Grass seed, 127-29, 138, 139
Green manuring, 33
Green matter, compost material, 57-58
Greensand, 43-44
Grinding of compost materials, 47
Ground covers, 134-36
Groundhogs, control of, 169
Ground rock fertilizers, 27, 39-40

Hair, compost material, 57
Hardpans, 6-8; formation by use of chemical fertilizers, 29-30
Heaps, compost, 59-64; success of, checkpoints to gauge, 60-61
Honey, darkening and crystallization, prevention of, 178; plants providing best, 177
House plants, 179-86; edible ornamentals, 185-86; eggshell placing method, 181; ferns, soil mixture used for, 180; fertilization for foliage plants, 183-84; fruits and vegetables, 184-85; light-grown, best light for, 181-82; mealy-bugs, 182-83; moisture retention, improving, 180; leaves turning light green, 179-81; warm and dark rooms, 182; water softeners, 182
Howard, Sir Albert, 2, 47
Hummingbirds, attracting to garden, 176
Humus, 1, 15, 19, 23-24, 46, 116; see also Compost

Indore method of composting, 46, 47
Insecticides, 142, 164-65, 176-77
Insects, control of, 1, 2, 140, 142-67; ants, 144, 155; aphids, 144, 155, 167; apple maggots, 162-64; bagworms, 144, 162; barn flies, 143; bean beetles, 144-45, 154; biological control of, 158-59; cabbage mag-

gots, 145; cankerworms, 94, 161; caterpillars, tent, 145-46; chinch bugs, 146, 154, 167; codling moths, 146-47; controlling with plants, 156-57; control program, 165; corn borers, 147, 155; corn earworms, 147, 154, 155; cucumber beetles, 147; cutworms, 147-48, 155; diseases, bees prone to, 177-78; earwigs, 148, 155; flies, 143, 153, 159; good and bad, 142-44; grasshoppers, 148-49, 154, 167; Japanese beetles, 149, 154; June bugs, 171; ladybird beetles, 167; ladybugs, 144, 151; leafhoppers, 149, 155; maggots, 149-50, 157, 162-64; May beetles, 171; mealybugs, 150, 182-83; mosquitoes, 150, 155, 159-60, 167; nematodes, 116, 150-51; peach borers, 151; pesticides, 142, 164-65; red spider mites, 151; root maggots, 151-52; scale, 152, 154, 161-62; slugs, 152, 155; snails, 152; sowbugs, 152, 154; squash bugs, 153, 154; tomato hornworms, 153, 155; trees, protection frcm, 161-64; trichogramma, 146, 166; war against, 142-43; white flies, 153; wintering of, 166-67; wireworms, 155, 156
Iron, 31
Irrigation ponds, 25

Japanese beetles, 149, 154
June bugs, 171

Kelp, 54-55
Kentucky bluegrass, 127, 138, 139
Kudzu, 24

Ladybird beetles, 167
Ladybugs, 144, 151
Lawns, 121-41; all-purpose, 122-23; ants and anthills in, 139-40; chinch bugs, 140; compost application, 52, 65-66, 134; disease susceptibility, conditions making, 137; drainage, proper, 137; elegant fine-bladed turf, 122; extra-sturdy working, 122; fertilizing, best time for, 133; grading, rate of slope in, 125; grass clippings on, 131; ground covers, 134-36; lime on, 133-34; moisture control, 137; mowing, 131; new, 125, 128; old, replanting, 125; powdery mildew, 137-38; problem areas, 128-29; seedbed for, preparing, 125-27; seed buying for, 127-28; seed mixture, purpose of using, 128; soil, testing before planting, 123, 124-25; soil sample for testing, preparing, 123-24; thatch, 136-37; trimming chores, easing, 132-33; types of, basic, 121-22; watering, 133; weed-free, maintaining, 140-41; zoysia, 129-31
Leaching, 37, 40
Lead arsenate, 72
Leaf curl of peach, 95
Leafhoppers, 149, 155
Leaves, compost process on, 51, 54
Legumes, 13-14, 118
Lespedeza, 13
Light for indoor plants, 181-82
Lime, 133-34
Linseed, 32
Listing, 25
Litter, importance of, 37
"Living Soil, The" (Balfour), 3

Maggots, 149-50; apple, 162-64; cabbage, 145; onion, 156-57; root, 151-52
Magnesium, 10, 31
Manganese, 13, 31, 41
Manure, 11, 33, 36-38, 118; application of, 36-37; compost making with, 54; crops benefited by application of, 36; fresh and rotted, difference between, 38; kind used, difference in, 37-40; loss of nutrients by, 37; storage of, 37
Manure worms, 74-75
May beetles, 171
Mealybugs, 150, 182-83
Merion bluegrass, 127, 138, 139

Mice, control of, 171-72
Microorganisms, soil, 20-23, 30, 38
Miscible oil spray solutions, 183
Mites, red spider, 151
Moisture control, lawns, 137
Moisture retention, improving, house plants, 180
Moles, control of, 169-70
Molner, Joseph G., 42
Mosquitoes, 150, 155, 159-60, 167
Moss, Spanish, compost material, 59
Mowing a lawn, 131
Mulch (mulching), 1, 25, 55, 78, 100-08, 192; advantages of, 101-02; coffee grounds as, 104; definition of, 100-02; disadvantages of, 102-03; fertilizing before mulching, 104; materials, 103-07; miscellaneous, 107-08; peanut hulls as, 104-05; peat, 107; pros and cons, 100-03; protection against winter weather, 107-08; quantity of, determining, 103-04; roses, 189; sawdust as, 105-06; shredded bark as, 107; tomato plants, 115
Municipal composting, advantages of, 70-71

Natural fertilizers, 28
Nematodes, 116, 150-51
Nettles, 19
New Zealand compost bin, 62-63
Nitrogen, 30, 31, 32-33, 42, 56, 57, 59; importance of, 34; percentage in kinds of manure, 38
Nitrogen fixation, 13
NPK analysis, 32

Oil sprays, 92-93, 183
Onion maggots, 156-57
Orange blossoms, 177
Organic, definition of, 3
Organic fertilizers, 32; chemical vs., 27-32; combining chemical and, 42; definition of, 28
Organic Gardening and Farming Magazine, 3

Ornamentals, edible, indoor growth of, 185-86

Pasteurization, 42-43
Pathogens, 42
Peach borers, 151
Peanut hulls as mulch, 104-05
Peanuts, 32
Pear trees, fire-blight on, combating, 94
Peas, 13
Peat, 11, 107
Perennials, 190-93; definition of, 190; grown from bulbs, 190; soil best for, 192-93; starting, best way for, 193; transplanting, 193; winter protection of, 193-94
Pesticides, 142, 164-65, 176-77
Pests, see Animal pests; Insects
pH, soil, 17-20, 30, 104; analysis of, 32; range for compost, 61
Phages, 20
Phosphate, 3, 30
Phosphorus, 31, 33, 56; percentage in kinds of manure, 38; significance of, 34; sources of, 35
Photoperiodism, 167
Physiological leaf roll, 114
Plantago, 19
Plantain, 19
Planting, seedlings of evergreens, 97-98; shrubs or trees grown in containers, 98; trees, 87-89; vegetables, 109-13
Plants, 179-96; annuals, 190-93; bees attracted to, 178; biennials, 190-93; birds attracted by, 175-76; controlling pests with, 156-57; forcing, 195-96; honey-making nectar from, 177; house, see House plants; ornamental, potted, feeding, 196; perennials, 190-93; protecting covers for, 194-95; retarding, 196; roses, 186-89; trees and shrubs, 88-96
Playgrounds, grass used on, 128-29
Plowing, 25
Poison ivy, eliminating, 79-80

INDEX

Potash, 31, 33, 54, 56; percentage in kinds of manure, 38

Potassium, 10, 31; role in plant nutrition, 35

Powdery mildew, 137-38, 187-89

Praying mantises, 145

Protein in sprouted seeds, 84

Protozoa, 22

Puddling, 11

Pyrethrum dust, 143

Quack grass, control of, 80-81

Rabbits, control of, 169, 171-72

Raccoons, control of, 172

Red spider mites, 151

Redwood shavings and sawdust, compost material, 69

Ring method, compost application, 67-68

Rock fertilizers, 27, 39-40

Rodale, J. I., 3

Rodale Press, Inc., 5

Root crops, manuring, 36

Root maggots, 151-52

Roses, 186-89; black spot, 186-87; hybrid and hedge or fencing types, 189; mulches for, 189; powdery mildew, 187-89; winter injury to, 189

"Roses for the Home" (U.S.D.A.), 186

Rotation of crops, 15-16

Rotenone dust, 143

Rough pigweed, 80

Rumex acetocella, 17

Rust, 116-17, 139

Rye, 11

Ryegrass, 127, 128

Sandy soils, 10

Sawdust as mulch, 105-06; disadvantages of, 106

Scale insects, 152, 154, 161-62

Scalicide, 95

Seaweed, 54-55

Sedges, 19

Seeds and seedlings, 77, 81-86; comparison of, with other foods, 86; cultivation of, 83-84; damp-off, 81-82; drying and storing, 86; evergreens, planting, 97-98; fruits and vegetables, soft-fleshed, saving, 86; germination, highest percentage of, 84-85; grass, 127-29, 138, 139; manure, used for next planting, 85-86; protein in, 84; stratification of, 82-83; vermiculite, use of, 82

Sewage waste, 27, 42

Shade trees, fertilizing, 98

Sheep sorrel, 17

Sheet compost, 14, 49, 51, 107

Shrubs, 87-99; compost application, 66-68; grown in containers, planting out of doors, 98; planting, 88-96; suffering from "winter drying," 98-99

Skunks, control of, 171

Sludge, 11, 32, 42

Slugs, 75-76, 152, 155

Smother mulch, 78

Snails, 152

Sodium, 10, 31

Sodium ions, 31

Soil, 1, 2, 6-25, 142; acid, 13; acid or alkaline condition of, determining, 17-18; aeration, 16-17; alkaline, 10-11; clay, 10; crop rotation, 15-16; erosion of, 11, 24-25; fertile, conditions making, 6-15; humus, 23-24; microorganisms, 20-23, 30, 38; mixture for potted ferns, 180; neutralization of, 13; perennials, best for, 192; pH, 17-20, 30, 104; quality of, determining, 31-32; samples for testing, preparing, 123-24; sandy, 10; structure of, improving, 14; testing, 18, 123, 124-25; unproductive, conditions making, 6-8

Soil builders, 26

Soil-testing methods, 31-32

Soil and Health Foundation, 3

Soil Association, England's, 3

Sorghum, 11

Sowbugs, 152, 154

INDEX

Soybeans, 32

Spanish moss, compost material, 59

Sprays (spraying), 1, 2, 74, 142, 143-44, 183; elimination of, in growing trees, 91-93

Squash bugs, 153, 154; control of, 117

Squash vine borer, control of, 117-18

Squirrels, control of, 170-71

Stink bugs, control of, 117

Stockwater ponds, 25

Stratification of seeds, 82-83

Straw stacks, decay of, hastening, 107

Stripcropping, 25

Striped cucumber beetles, control of, 117

Stubble mulching, 25

Sugar beets, 11

Sulfates, 30

Sweep plowing, 25

Sweet clover, 19, 33

Sykes, Friend, 4

Tanglefoot compound, 161

Tent caterpillars, 145-46, 162

Thatch, 136-37

Timothy, 128

Tomato hornworms, 153, 155

Tomato plants, blossom end rot of fruit, 115; blossoms drop off without setting fruit, 114; crack-of fruits, causes of, 114; curl of leaves on, 114; mulching, 115; poor growth of, avoiding, 118; rules for growers, general, 115-16; staking of, 115

Trace elements, 31, 40-41

Traffic areas, grass used on, 128-29

Trees, 87-99; canker worms, discouraging, 94; chemical sprays in growing, 91-93; compost application, 66-67; drainage for planting, proper, 96-97; evergreens, see Evergreens; fire-blight on pear, combatting, 94; fruit, see Fruit trees; grown in containers, planting out of doors, 98; leaf curl of peach, 95; planting, best method for, 87-89; protecting from insects, 161-64; shade, fertilizing, 98; whitewashing trunks, advantages of, 95; woodpeckers and, 176; wrapping trunks of newly transplanted, 95

Trichogramma, 146, 166

Trik-o, 161, 162

Vegetable garden, compost application, 65; sawdust mulch in, 105

Vegetable meals, 32

Vegetables, 109-20; bean rust, overcoming crop loss to, 116-17; carrots, getting well-formed roots, 116; fertilization of, 65, 118-20; indoor growing of, 184-85; not to be planted together, 111; planting, 109-13; planting directions, 112-13; planting combinations, 109-11; planting table and requirements, 112-13; problems, 114-18; squash bugs, control of, 117; squash vine borers, control of, 117-18; striped cucumber beetles, control of, 117; time of year to plant, 111, 112-13; tomato plants, see Tomato plants.

Velvet peas, 13

Vermiculite, 82

Vetch, 13

Violets, 19

Viruses, 20

Water softeners for indoor plants, 182

Weeds, 77; eliminating, 78-81

White flies, 153

Whitewashing tree trunks, 95

Windbreaks, 25

Winter drying, shrubs suffering from, 98-99

Wireworms, 155, 156

Wood ashes, compost material, 56

Woodpeckers, trees injured by, 176

Wrapping trunks of newly transplanted trees, 95

Yeasts, 21

Zoysia grass lawns, advantages of, 129-30; disadvantages of, 130-31